# FLORA OF TROPICAL EAST AFRICA

## FLACOURTIACEAE

H. Sleumer
(Rijksherbarium, Leiden)

Trees or shrubs, sometimes with spines on trunk and branches. Leaves spirally arranged, or often distichous, simple, entire, crenate or serrate (crenations mostly glandular), sometimes with pellucid dots and/or lines; lateral nerves pinnate or rarely several from the base; petioles often thickened at the base and (or) the apex; stipules generally present, rarely large and persistent. Inflorescences subterminal or mostly axillary, rarely on the trunk or very rarely on the midrib of the leaves; flowers solitary, or mostly in fascicles, racemes or panicles, apparently essentially cymose, bisexual or unisexual (polygamous, monoecious, or dioecious), regular, 3–poly-merous, spiral or cyclic. Pedicels often articulated near the base. Sepals (2–)3–7 (rarely more), mostly persistent, sometimes accrescent, imbricate or valvate, free or connate at the base into a calyx-tube. Petals 3–8 (rarely more), free, imbricate or valvate, mostly alternating with the sepals and caducous, sometimes persistent and accrescent, often inserted on the margin of a hypogynous or almost perigynous disk, or absent. Receptacle often depressed in the centre, often with appendages such as an extrastaminal disk or disk-lobes, free glands between the stamens, or staminode-like scales inserted on the inner side of the base of the petals, or with true, mostly barbate staminodes which alternate with the stamens. Stamens 5 to numerous, hypogynous, mostly free, or sometimes connate at the base with the staminodes; anthers with 2 thecae, these longitudinally dehiscent, or very rarely opening by apical pores or very short slits. Ovary usually free, rarely semi-inferior, unilocular, with (2–)3–5(–8) parietal placentas; ovules generally numerous, anatropous; styles 1–10, free or connate; stigma sessile. Fruit a fleshy or dry berry or a capsule, rarely a drupe, 1–many-seeded. Seeds sometimes arillate, with abundant endosperm; embryo straight; cotyledon mostly broad, foliaceous.

A pantropical family represented by 21 genera and 43 species in East Africa, out of a world total of about 85 genera and approximately 1100 species.

Inner perianth-segments (petals) present, sometimes hardly differentiated from the outer ones (sepals):
Perianth-segments either hardly differentiated from each other, or, if so, the inner ones more numerous than the outer ones and fixed independently in relation to the latter:
Perianth-segments spirally arranged, each of the inner ones (petals) provided with a basal internal scale:
Outer perianth-segments (sepals) free and about half the size of the inner ones

1

(petals); racemes subspicate, rather
short . . . . . . 1. **Rawsonia**
Outer perianth-segments (sepals) connate
in their lower third and ± the same
size as the inner ones (or petals);
racemes slender, elongate . . . 2. **Dasylepis**
Perianth-segments cyclically arranged, the
inner ones (petals) without a basal
internal scale:
Fruit winged at least in its upper part:
Leaves deciduous; flowers bisexual; ovary
slightly stipitate; fruit 4-winged . 3. **Peterodendron**
Leaves persistent; flowers andromonoeci-
ous; ovary sessile; fruit 6-winged . 4. **Grandidiera**
Fruit not winged:
Sepals valvate; fruit echinate with branching soft bristles or spines . . . 5. **Buchnerodendron**
Sepals imbricate; fruit smooth, or with
simple bristles or spines:
Branches spiny; fruit indehiscent . 6. **Oncoba**
Branches unarmed; fruit dehiscent by
3–8 valves:
Fruit with a thick woody pericarp;
petiole not swollen distally . 7. **Xylotheca**
Capsule with a leathery (sometimes
fibrous) pericarp; petiole swollen distally:
Flowers from leafless axils or old
wood, large; petals 18–35 mm.
long; fruit with numerous
seeds . . . . . 8. **Caloncoba**
Flowers from the axils of upper
leaves, rather small; petals
7–23 mm. long; fruit 1–3-
seeded . . . . 9. **Lindackeria**
Perianth-segments showing a generally equal
number of distinct sepals and petals:
Inflorescences epiphyllous . . . . 10. **Phylloclinium**
Inflorescences axillary, terminal or cauline:
Petals each provided internally with a
fleshy scale; anthers opening by pores
or very short slits . . . . 11. **Kiggelaria**
Petals without such a scale; anthers opening by normal elongate pores:
Stamens indefinite and irregularly arranged; fruit a fleshy berry . . 12. **Scolopia**
Stamens solitary or in fascicles opposite
each petal; fruit dry or somewhat
fleshy, sometimes dehiscent:
Style 1 . . . . . 13. **Gerrardina**
Styles 2–6:
Flowers dioecious; leaves with 5–7
basal nerves; seeds with a red
aril; testa tessellate . . . 14. **Trimeria**
Flowers bisexual; leaves penni-
nerved; seeds without an aril . 15. **Homalium**

Inner perianth-segments (petals) absent:

Stamens connate at the base with the alternating disk-glands, and on the edge of a ± depressed receptacle; leaves (at least initially) with marked pellucid dots and/or lines . 16. **Casearia**

Stamens free (as are the disk-glands, if present); leaves very rarely with very fine pellucid dots (*Dovyalis*), generally without pellucid dots or lines:

Stamens in fascicles which alternate with the sepals; fruit a capsule . . . . 17. **Bivinia**

Stamens irregularly arranged; fruit a berry or a berry-like drupe:

Ovary 1-locular; spines usually absent:

Disk-glands present; style 2–4-fid in its upper part; berry reddish-yellowish . 18. **Ludia***

Disk-glands absent; style very short; stigma subsessile, almost discoid; berry white . . . . . 19. **Aphloia**

Ovary falsely 3–8-locular; spines generally present:

Fruit a berry-like drupe with 1-seeded pyrenes in pairs one above the other; stamens surrounded by a ring of disk-glands . . . . . 20. **Flacourtia**

Fruit a berry with 1 or 2 seeds; stamens alternating with the disk-glands . 21. **Dovyalis**

## 1. RAWSONIA

Harv. & Sond. in Fl. Cap. 1: 67 (1860); Gilg in E. & P. Pf., ed. 2, 21: 394 (1925)

Shrubs or trees. Leaves persistent, alternate, rather leathery, entire to crenate or spinulose-serrate, petiolate, penninerved, stipulate. Flowers small, white or pinkish, polygamous, i.e. the lower ones in the same inflorescence generally ♂, the upper 1–3 ♀ or ⚥, in axillary spike-like racemes. Perianth spirally arranged, consisting of 4–6 free, imbricate, concave and unequal sepals (the outer ones the smallest, the inner ones gradually larger), and 4–8 petals, similar to the sepals but successively larger towards the centre of the flower, and each bearing a fleshy glabrous or hairy scale-like basal appendage. Stamens numerous, biseriate in the ⚥, multiseriate in the ♂ flowers, the inner row arising from the receptacle, the outer row attached at the base of the " scales "; filaments short; anthers subsagittate. Ovary on a concave receptacle, 1-locular, with 3–5 multi-ovulate placentas; style very short; stigmas 3–4(–5), radiate. Fruit berry-like when first ripe, but tardily longitudinally dehiscent when dry into 3–5 sections; pericarp hard. Seeds few, subglobose-angular; aril pergamaceous.

A genus of 2 species, both in Africa and also represented in East Africa.

Spikes with ± densely and regularly arranged flowers, and 2–3(–6) mm. long peduncle, i.e. set with flowers

---

* Species of *Scolopia* with small caducus petals may sometimes be confused here.

to almost the base of the stoutish rhachis, (0·5–)1–
2·5(–4) cm. long in all . . . . . .  1. *R. lucida*
Spikes with ± laxly and irregularly arranged flowers and
(1·2–)2–4(–5) cm. long slender peduncle, 4·5–7 cm.
long in all . . . . . . . .  2. *R. reticulata*

1. **R. lucida** *Harv. & Sond.* in Fl. Cap. 1: 67 (1860); Gilg in E.J. 40:
449 (1908); V.E. 3 (2): 559 (1921); Wild in F.Z. 1: 262, t. 42/A (1960);
K.T.S.: 227, fig. 44 (1961); F.F.N.R.: 266 (1962); Bamps in F.C.B.,
Flacourt.: 4 (1968). Type: South Africa, Natal, Durban, *Sanderson* 118
(TCD, holo., Kew Neg. 12163 !, K, iso. !)

Shrub (rarely subscandent) or tree, rarely up to 20 m. tall, trunk up to
50 cm. across; bark grey, peeling in rounded scales. Leaf-blades lanceolate
to oblanceolate, rarely oblong or oblong-obovate, glabrous, glossy, subcoriaceous to coriaceous, entire or generally shallowly to (particularly on
coppice shoots) more deeply spinulose-serrate, 7–16(–20) cm. long, (2–)2·5–
6(–8·5) cm. broad; nerves 6–9 pairs, slightly raised on both faces, as is the
rather dense reticulation; petiole 5–12(–15) mm. long, glabrescent; stipules
narrowly oblong, ciliate, 5–7(–10) mm. long, caducous. Flowers in axillary,
spike-like, densely flowered racemes; rhachis stoutish, (0·5–)1–2·5(–4) cm.
long, its lowest 2–3(–6) mm. without flowers; pedicels robust, 1(–2) mm.
long. Sepals unequal, suborbicular to elliptic, (2–)3–4(–5) mm. long,
1·5–3·5 mm. broad, ciliolate. Petals similar to the sepals, ± twice their
size, with a fleshy pubescent elliptic or bilobed basal scaly appendage.
Stamens 25–30. Ovary glabrous or slightly pubescent; style 1 or 2 mm.
long; stigmas short. Fruit globular, rarely ovoid, yellow, finally reddish,
the style persistent as a short point, a little fleshy and berry-like at first,
tardily dehiscent into 4 or 5 longitudinal sections when dry, 2–2·5(–4) cm.
across; pericarp (2–)3–6 mm. thick. Seeds few, ovoid to subglobose, 6–10 mm.
across. Fig. 1/1–12.

UGANDA. Bunyoro District: Budongo Forest, Apr. 1938, *Eggeling* 3589 !; Toro District:
Dura R., July 1951, *Osmaston* 1008 !; Busoga District: Butembe-Bunya, Mpomode
Hill, Sept. 1952, *G. H. Wood* 363 !
KENYA. Nairobi District: Karura Forest, Mar. 1958, *Verdcourt* 2141 !; Masai District:
Emali Hill, Mar. 1940, *V. G. van Someren* 130 !; Kwale District: Shimba Hills, Jan.
1964, *Verdcourt* 3926 !
TANZANIA. Mwanza District: Samina, Jan. 1953, *Procter* 121 !; Lushoto District:
Kwai valley, Apr. 1953, *Drummond & Hemsley* 2238 !; Iringa District: Mufindi,
Nov. 1947, *Brenan & Greenway* 8259 !; Pemba I., Ngezi forest, Feb. 1929, *Greenway*
1484 !
DISTR. U2, 3; K4–7; T1–3, 6–8; P; Somali Republic, Sudan, Angola, Zaire, Malawi,
Zambia, Rhodesia, Mozambique, Swaziland, South Africa (Transvaal, Natal, Cape
Province southwards to Port St. John)
HAB. Understorey and shrub layer of lowland and upland rain-forest, dry evergreen
forest, semi-swamp and riverine forest; 50–1900 m.

SYN. *Oncoba spinidens* Hiern, Cat. Afr. Pl. Welw. 1: 39 (1896). Type: Angola,
Cuanza Norte, Pungo Andongo, *Welwitsch* 886 (LISU, holo., BM, K, iso. !)
*Rawsonia ugandensis* Dawe & Sprague in J.L.S. 37: 500 (1906); I.T.U., ed. 2:
150 (1952). Type: Uganda, Bunyoro District, Bugoma Forest, *Dawe* 946
(K, holo. !)
*R. schlechteri* Gilg in E.J. 40: 449 (1908); V.E. 3 (2): 559 (1921); T.T.C.L.: 235
(1949). Types: Tanzania, Lushoto District, Derema, *Scheffler* 154 (B, syn. †,
BM, BR, K, P, isosyn. !) & Msituni, *Scheffler* 231 (B, syn. †) & Amani,
*Zimmermann* in Herb. *Amani* 1011 (B, syn. †) & Malawi, Blantyre, *Buchanan*
293 [Herb. *Wood* 6886] (B, syn. †, BM, K, isosyn. !)
*R. usambarensis* Engl. & Gilg in E.J. 40: 449 (1908); V.E. 3 (2): 559 (1921);
T.S.K.: 23 (1936); T.T.C.L.: 235 (1949). Types: Kenya, *Scott Elliott* 87 (B,
syn. †, K, isosyn. !) & Tanzania, Mwanza, Ukerewe I., *Holtz* 1566, 1578 &

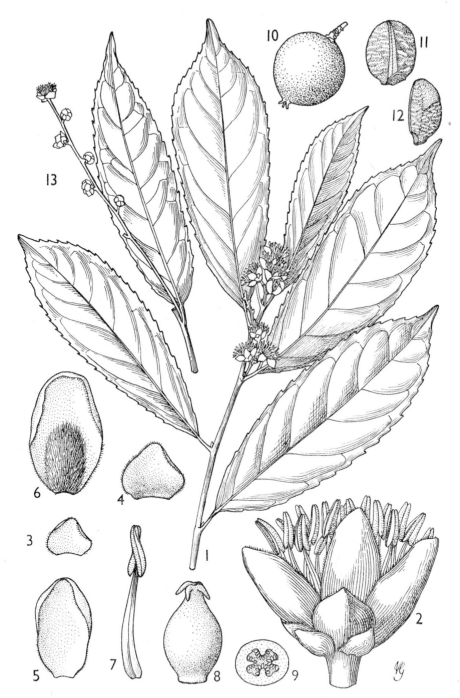

FIG. 1. *RAWSONIA LUCIDA*—**1**, flowering branch, × ⅔; **2**, flower, × 6; **3–5**, sepals × 6; **6**, petal, × 6; **7**, stamen, × 6; **8**, pistil, × 6; **9**, transverse section of ovary, × 6; **10**, fruit, × 1; **11, 12**, seed, two views, × 2.   *R. RETICULATA*—**13**, part of flowering branch, × ⅔.   1, from *Drummond & Hemsley* 3409; 2–9, from *Lewis* 226; 10, 11, 12, from *Williams* 361; 13, from *Drummond & Hemsley* 1748. Drawn by Victoria Goaman.

*Uhlig* V-56 & Lushoto District, Kwai, *Eick* 33, 37, 76 & *Albers* 366 (all B, syn. †)
*R. transjubensis* Chiov., Fl. Somala 2: 12, fig. 3 (1932); E.P.A.: 596 (1959).
    Type: Somali Republic (S.), Transjuba, Ola–Uager, *Senni* 38 (FI, holo.!)
*R. spinidens* (Hiern) Mendonça & Sleumer in C.F.A. 1: 79 (1937)

2. **R. reticulata** *Gilg* in E.J. 30: 357, fig. (1901) & 40: 449 (1908);
V.E. 3 (2): 559, fig. 247 (1921); Gilg in E. & P. Pf., ed. 2, 21: 394, fig. 165
(1925). Type: Tanzania, Njombe District, Kipengere Range, *Goetze* 983
(B, holo. †, BR, iso.!)

Shrub or small tree, rarely up to 10 m. tall, much branched; bark
smooth, grey to brownish, flaking off in small irregular patches. Branchlets
reddish-brownish, tips glabrous. Leaf-blades oblong or oblanceolate,
sometimes elliptic-oblong, apex shortly acuminate, base broadly cuneate to
rounded, subcoriaceous to coriaceous, glabrous, in most characters similar
to those of *R. lucida*, though the dentation is often less sharp, (7–)10–
14 cm. long, 2·5–6·5 cm. broad; nerves 6–9 pairs, raised on both faces as is
the lax reticulation; petiole 5–7 mm. long. Spikes peduncled, i.e. without
flowers for their lowest (1·2–)2–4 cm., upper part of the rhachis slender,
with 4–9 laxly or distantly and rather irregularly arranged flowers, glabrous,
4·5–7 cm. long including the peduncle; pedicels rather slender, 0·5–1 mm.
long. Flowers very similar to those of *R. lucida*, generally a little
smaller. Stamens 14–18. Ovary glabrous or slightly pubescent; style very
short; stigmas 3 or 4. Fruit ovoid-subglobose, glabrous, 2–3 cm. long, 1·3–
2 cm. across; pericarp firmly coriaceous, possibly thicker than that of
*R. lucida* in the mature state; pedicel 2–3 mm. long. Seeds 3–8. Fig. 1/13.

TANZANIA. Lushoto District: Magamba Pass, Nov. 1931, *Wigg* 28!; Mpwapwa
    District: Kiboriani Mts., Oct. 1938, *Greenway* 5797!; Morogoro District: Bunduki,
    Mar. 1953, *Semsei* 1100!
DISTR. T3, 5–7; Malawi
HAB. In second storey layer of upland rain-forest; 1200–2400 m.

SYN. *R. uluguruensis* Sleumer in N.B.G.B. 11: 1077 (1934). Type: Tanzania, Uluguru
    Mts., Mt. Lupanga, *Schlieben* 2965 (B, holo. †, B, BM, BR, M, P, iso.!)
    *Dasylepis burtt-davyi* Edlin in K.B. 1935: 255 (1935); Wild in F.Z. 1: 263,
    t. 42/B (1960). Type: Malawi, Mt. Mlanje, *Burtt Davy* 22043 (FHO, holo.!,
    K, iso.!)

## 2. DASYLEPIS

Oliv. in J.L.S. 9: 170 (1865); Gilg in E. & P. Pf., ed. 2, 21: 394 (1925);
Sleumer in E.J. 92: 554 (1972)

Shrubs or trees. Leaves persistent, alternate, entire, dentate or serrate,
glabrous, penninerved; stipules caducous. Racemes axillary, solitary;
pedicels subtended by several minute bracts. Flowers bisexual or poly-
gamous-monoecious, i.e. the lower ones in the same raceme often ♂ only.
Sepals 8–12, spirally arranged and subequal; outer 3–5 coherent below,
thin to chartaceous and generally reddish or pinkish; inner 5–7 ones almost
free, membranous and whitish, each with a small thick hairy scale adhering
to the base on the inner side. Stamens 16–30; filaments filiform; anthers
linear-oblong, subsagittate, basifixed. Ovary with 2–4 multi-ovulate placentas; style short, divided distally into (2), 3 or 4 stigmatic branches. Fruit
capsular, finally splitting into 3 valves; pericarp coriaceous. Seeds 1–5,
obtusely ovoid-angular; testa hard; endosperm copious; cotyledons flat.

A genus of 6 species in tropical Africa, 3 of which are represented in East Africa.

Ovary tomentose; fruit pubescent, tardily glabrescent .   1. *D. eggelingii*
Ovary and fruit glabrous (or practically so) from the
  beginning:
Rhachis stoutish, 1·5–2 mm. across; fruit 2·5–3 cm.
  across; pericarp ± 5 mm. thick      .      .      .   2. *D. racemosa*
Rhachis slender, 0·5–1 mm. across; fruit 1·6–2 cm.
  across; pericarp 1·5–2·5 mm. thick  .      .      .   3. *D. integra*

1. **D. eggelingii** *Gillett* in K.B. 5: 341 (1951); I.T.U., ed. 2: 146 (1952);
Bamps in F.C.B., Flacourt.: 8 (1968); Sleumer in E.J. 92: 555 (1972).
Type: Uganda, Ankole District, Kalinzu Forest, *Eggeling* 3710 (K, holo.!,
BM, BR, iso.!)

Shrub or tree, glabrous, up to 10 m. tall; bark thin, dark red-brown.
Branchlets brownish-yellowish, sparsely lenticellate.  Leaf-blades oblanceolate or oblong, apex acuminate, tip bluntish, base cuneate to rounded, thin
to coriaceous, regularly serrate, 10–19 cm. long, 3–7 cm. broad; nerves
6–8(–9) pairs, prominent on both faces, as is the rather dense though much
finer reticulation; petiole 5–12 mm. long. Racemes pendulous, 5–16-flowered;
rhachis slender, glabrous, 6–15(–18) cm. long, no flowers in the lower ⅓ to
½; pedicels slender, 5–10(–15) mm. long at anthesis. Sepals 10–12, white to
pinkish, the outer 3–5 elliptic and 6–7 mm. long, 5–6 mm. broad, the inner
ones elliptic-ovate to -obovate, ± 7 mm. long and 2·5–5 mm. broad, their
pubescent scale ± 2 mm. long.  Stamens 25–30; filaments ± 5 mm. long;
anthers 3 mm. long.  Ovary shortly tomentose to almost woolly; style ±
1 mm. long, with 3 or 4 very short stigmatic branches.  Fruit subglobular,
1·5–1·8 cm. across, densely pubescent, finally purple-red; pericarp 0·5–1 mm.
thick.  Seeds ± 3, ± 8 mm. long, 7 mm. broad.

UGANDA.  Toro District: Fort Portal Forest Reserve, July 1960, *Paulo* 579!; Masaka
    District: Sango Bay Forest, July 1950, *G. H. Wood* 758!; Mengo District: Mabira
    Forest, Mar. 1907, *Ussher* 2!
DISTR.  U2, 4; Zaire
HAB.  Understorey shrub or tree in rain-forest; 1400–1525 m.

SYN.  *D. lebrunii* Evrard in B.S.B.B. 86: 8 (1953).  Type: Zaire, Irumu–Mombasa,
    *Lebrun* 4137 (BR, holo.!)

2. **D. racemosa** *Oliv.* in J.L.S. 9: 170 (1865) & F.T.A. 1: 123 (1868) & in
Hook., Ic. Pl. 11, t. 1029 (1868); V.E. 3 (2): 560 (1921); F.W.T.A., ed. 2, 1:
186 (1954); Bamps in F.C.B., Flacourt.: 5 (1968); Sleumer in E.J. 92:
556 (1972).  Type: Cameroun, Mt. Cameroon, *Mann* 2149 (K, holo.!, P,
iso.!)

Tree up to 20 m. tall, glabrous; trunk up to 45 cm. across, often fluted;
bark thin, orange-brown, rough, flaking in patches.  Branchlets greyish-
brownish, laxly set with lenticels.  Leaf-blades elliptic to oblanceolate or
obovate, apex shortly obtusely acuminate, base cuneate to rounded, firmly
chartaceous to subcoriaceous, ± remotely serrate, (10–)12–22 cm. long,
(4–)6–9 cm. broad; nerves (4), 5 or 6 pairs, their looping not much obvious;
reticulation rather dense, raised on both faces; petiole 7–15(–20) mm. long.
Racemes obliquely erect, laxly many-flowered, glabrous; rhachis stoutish,
5–11 cm. long, 1·5–2 mm. across, with the flowers generally down almost to
the base; bracts broadly ovate, ciliolate, ± 1 mm. long, 2 mm. broad;
pedicels rather slender, 6–10 mm. long at anthesis. Sepals white, sometimes cream or pinkish, glabrous, outer ± 4 elliptic and 5–7 mm. long,
2–2·5 mm. broad, inner ± 7 obovate and ± 5 mm. long, 2–3 mm. broad,
their pubescent scale ± 2 mm. long.  Stamens ± 30; filaments 3–4 mm.

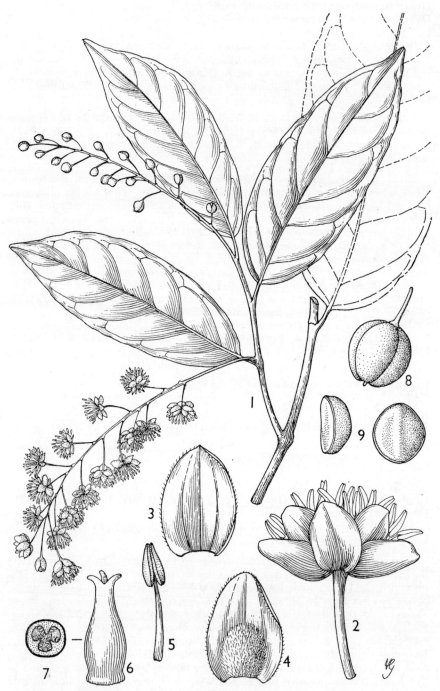

FIG. 2. *DASYLEPIS INTEGRA*—**1**, flowering branch, × ⅔; **2**, flower, × 4; **3**, sepal, × 6; **4**, petal, × 6; **5**, stamen, × 6; **6**, pistil, × 12; **7**, transverse section of ovary, × 12; **8**, fruit, × 1½; **9**, seed, × 1½.  1, from *Drummond & Hemsley* 4341; 2–7, from *Drummond & Hemsley* 1402; 8, 9, from *Faden* 71/19. Drawn by Victoria Goaman.

long; anthers 2–3 mm. long. Ovary glabrous; style ± 2 mm. long, divided distally into 2, (3) or 4 very short stigmatic branches. Fruit subglobular, slightly apiculate, reddish, finally splitting in 3 valves, 2·5–3 cm. across; pericarp ± 5 mm. thick. Seeds few, ± 10 mm. long, 8 mm. broad.

UGANDA. Kigezi District: Kayonza Forest Reserve, Feb. 1945, *Greenway & Eggeling* 7101! & Impenetrable Forest, Apr. 1948, *Purseglove* 2666 & Rujezi (?Rugezi), Aug. 1948, *St. Clair-Thompson* 2502!
DISTR. **U**2; Nigeria, Cameroun, Gabon, Zaire, Rwanda
HAB. Understorey tree in upland rain-forest; 1280–2440 m.

SYN. [*D. leptophylla* sensu I.T.U., ed. 2: 146 (1952), *non* Gilg]

3. **D. integra** *Warb.* in P.O.A. C: 277 (1895); T.S.K.: 22 (1936); T.T.C.L.: 230 (1949); K.T.S.: 224 (1961); Sleumer in E.J. 92: 557 (1972). Type: Tanzania, Lushoto District, Gonja Forest, *Holst* 4262 (B, holo. †, BM, HBG, K, M, P, W, iso.!)

Shrub or tree, up to 12(–35) m. tall; trunk slender; bark brownish-greyish, smooth, flaking in old specimens. Leaf-blades elliptic to lanceolate or oblong, rarely ovate-elliptic, apex shortly obtusely acuminate, base cuneate to almost rounded, chartaceous to subcoriaceous, reddish in flush, entire to rather remotely serrate-dentate, 8–16(–18) cm. long, 2·5–6(–8) cm. broad; nerves (5–)6–8(–9) pairs, indistinctly looping, raised mainly beneath, reticulation ± dense, slightly prominent on both faces, though more obvious beneath; petiole 5–7(–10) mm. long. Racemes ± drooping, laxly many-flowered, glabrous; rhachis (4–)6–12 cm. long, ± 1 mm. in diameter at base, without flowers in its lower part; bracts minute; pedicels slender, 5–8(very rarely –20) mm. long. Sepals 10–12, glabrous, the outer 3 or 4 pinkish to reddish and ± 5 mm. long, 4 mm. broad, the inner 7 or 8 slightly longer and more whitish, their pubescent scale 1·5–2 mm. long. Stamens 15–20; filaments 3–4 mm. long; anthers ± 2 mm. long. Ovary glabrous, tapering to a very short style with 3 stigmatic branches. Fruit subglobose, a little rostrate, with 3 or 4 longitudinal grooves, finally 3-valvate, pale purple, 1·5–2 cm. across; pericarp 1·5–2·5 mm. thick. Seeds 1–2(–5), 8–9 mm. long, 6–7 mm. broad. Fig. 2.

KENYA. Teita District: Teita Hills, Sept. 1953, *Drummond & Hemsley* 4341!
TANZANIA. Mbulu District: Nou Forest Reserve, Aug. 1958, *Manolo* 176!; Pare District: Pare Mts., Mtonto, July 1942, *Greenway* 6524!; Lushoto District: Mkusu Valley, Apr. 1953, *Drummond & Hemsley* 2218!
DISTR. **K**7; **T**2, 3, 6; not known elsewhere
HAB. Shrub or tree in understorey of rain-forest; 900–2200 m.

SYN. *D. leptophylla* Gilg in E.J. 40: 450 (1908); V.E. 3 (2): 561 (1921); T.T.C.L.: 230 (1949). Type: Tanzania, Lushoto District, Kwai–Gare, *Engler* 1212 (B, lecto. †, EA, isolecto.)
*D. sp.* sensu T.S.K.: 22 (1936). Based on a specimen from Kenya, Teita Hills

## 3. PETERODENDRON

Sleumer in N.B.G.B. 13: 357 (1936)

Shrubs or small trees. Leaves deciduous, alternate, petiolate, penni-nerved, stipulate. Flowers bisexual, solitary in the axils of the upper leaves, on a long peduncle articulated halfway. Sepals 3, imbricate, thin. Petals 8, obovate, clawed, imbricate, very thin, expanded during anthesis. Stamens 45–50, multiseriate, free; filaments filiform; anthers elongate-oblong, basi-fixed. Ovary 1-locular, shortly stipitate, with 4 lateral rapidly growing wings, with 2 multi-ovulate placentas; style filiform; stigma punctiform. Capsule subglobular, the upper part provided with 4 ascending or ±

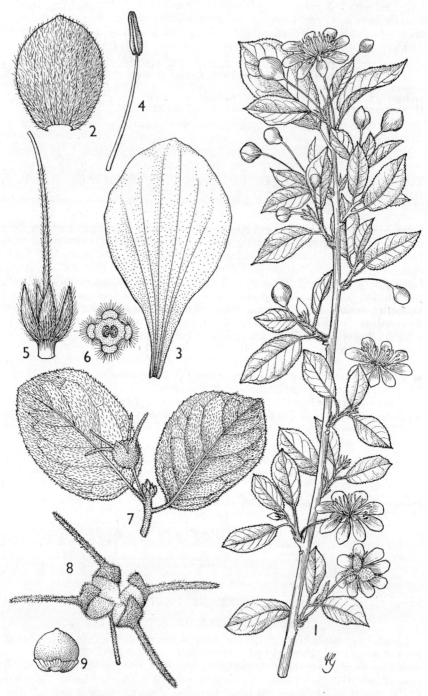

FIG. 3. *PETERODENDRON OVATUM*—**1**, flowering branch, × ⅔; **2**, sepal, × 4; **3**, petal, × 4; **4**, stamen, × 4; **5**, pistil, × 4; **6**, transverse section of ovary, × 4; **7**, part of fruiting branch, × ⅔; **8**, dehisced capsule, × 1⅗; **9**, seed, × 1⅗. 1, from *Greenway & Kanuri* 14788; 2–6, from *B. D. Burtt* 5400; 7–9, from *Renvoize & Abdallah* 2304. Drawn by Victoria Goaman.

vertical, almost linear subacute wings; pericarp crustaceous. Seeds 1 or 2, globular, with a red shallowly cup-shaped aril.

A monotypic genus restricted to Tanzania.

**P. ovatum** (*Sleumer*) *Sleumer* in N.B.G.B. 13: 358 (1936); T.T.C.L.: 234 (1949). Type: Tanzania, Kilosa District, Elphon's Pass to Ruaha R., *Troll* 5069 (B, holo. †)

Shrub or small tree, 1·5–2·5(–5) m. tall; bark brown, lenticellate. Tips of branchlets velvety. Leaf-blades ovate or oblong-ovate, apex shortly acuminate, base rounded to subcordate, thin to papyraceous, velvety when young, subglabrous above and pubescent beneath when mature, denticulate, 5–8 cm. long, 4–7 cm. broad; lateral nerves 5–7 curved-ascending pairs, raised beneath, reticulation rather dense; petiole (1–)1·5–3 cm. long; stipules subulate, caducous, ± 5 × 1 mm. Flowers on peduncles (1·5–)2–2·5 cm. long, ± pendulous. Sepals suborbicular, pubescent, ± 6 mm. in diameter. Petals obovate-spathulate, white, ± 1·5 cm. long, 6–7 mm. broad. Capsule subglobose, pubescent, ± 1 cm. across, with 4 linear subacute wings ± 1·5 cm. × 1 mm., splitting into 4 patent valves. Seeds 1 or 2, subglobular, blackish, glossy, 5–6 mm. in diameter, the base provided with a red aril. Fig. 3.

TANZANIA. Kondoa District: Kondoa-Irangi, Dec. 1935, *B. D. Burtt* 5328!; Kilosa District: Kidete, Dec. 1925, *Peter* 32741!; Iringa District: Ruaha National Park, Kimiramatonge Hill, July 1969, *Greenway* 14215!
DISTR. T5–7; not known elsewhere
HAB. Deciduous bushland and thicket; 700–1300 m.

SYN. *Poggea ovata* Sleumer in N.B.G.B. 12: 475 (1935)


## 4. GRANDIDIERA

Jaub. in Bull. Soc. Bot. Fr. 13: 467 (1866); Gilg in E. & P. Pf., ed. 2, 21: 399, fig. 170 (1925)

Shrubs or trees. Leaves persistent, alternate, petiolate, entire or undulate, penninerved, stipulate. Flowers in short axillary spike-like racemes; pedicels articulate at the base, elongate in the terminal bisexual flower, short in the lower ♂ ones; subtending bracts several, membranous, small. Sepals 3, imbricate, small. Petals 5–7, imbricate, small. Male flowers: stamens numerous; filaments filiform; anthers subovate-oblong, dorsifixed near the base, longitudinally dehiscent; rudiment of ovary 0. Hermaphrodite flowers: sepals and petals as in the ♂ ones, though slightly larger; stamens less numerous (35–50); ovary unilocular, sessile, with 2–4 multi-ovulate placentas, shortly 4–6(–8)-winged; style short; stigmas (2–)3(–4), divergent, reflexed. Fruit globular, woody, tardily dehiscent, with 4–6(–8) short membranous crenate wings. Seeds numerous, ovoid, small.

A monotypic genus restricted to eastern Africa.

**G. boivinii** *Jaub.* in Bull. Soc. Bot. Fr. 13: 467 (1866); Oliv. in F.T.A. 1: 119, in obs. (1868); Warb. in P.O.A. A: 33, & C: 277 (1895); Gilg in E.J. 40: 453 (1908); V.E. 3 (2): 563 (1921); Gilg in E. & P. Pf., ed. 2, 21: 399, fig. 170 (1925); T.T.C.L.: 232 (1949); K.T.S.: 226 (1961); Paiva in Bol. Soc. Brot., sér. 2, 40: 264, t. 1 (1966); Exell in F.Z. 3: 141 (1970). Types: Zanzibar, *Grandidier* 28 (P, lecto.!) & 1848, *Boivin* & Kenya, Mombasa, 1848, *Boivin* (both P, syn.!)

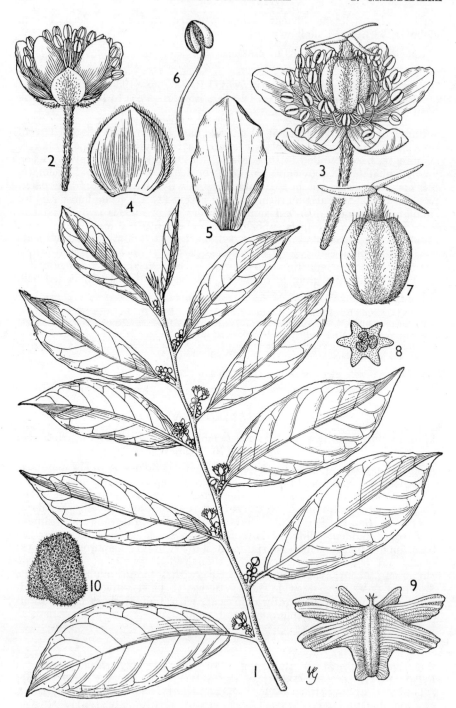

FIG. 4. *GRANDIDIERA BOIVINII*—**1**, flowering branch, × ⅔; **2**, male flower, × 4; **3**, hermaphrodite flower, × 6; **4**, sepal, × 8; **5**, petal, × 8; **6**, stamen, × 12; **7**, pistil, × 8; **8**, transverse section of ovary, × 8; **9**, fruit, × 1½; **10**, seed, × 4. 1, 3–8, from *Drummond & Hemsley* 3878; 2, from *Faulkner* 531; 9, from *Faulkner* 18541; 10, from *Verdcourt* 198. Drawn by Victoria Goaman.

Shrub or tree, 2–6(–10) m. tall.  Branchlets with short ± appressed hairs at tips, glabrescent below.  Leaf-blades oblanceolate to obovate, apex elongately and acutely acuminate, often a little curved, base gradually cuneate, papyraceous, glabrous except some hairs on midrib and nerves beneath, entire or shallowly undulate, 8–25 cm. long, 3·5–5(–7·5) cm. broad; lateral nerves 8–12 pairs, curved-ascending, looping before the edge, a little prominent beneath; petiole pubescent initially, 0·7–2(–2·5) cm. long; stipules subulate, pubescent, finally caducous, 6–7 mm. long.  Racemes spike-like from the upper axils, few-flowered, 0·5–2·5 cm. long, shortly hairy all over; pedicel of the terminal bisexual flower 5(–7) mm., of the lower ♂ ones 1–2 mm.  Sepals suborbicular, pale green, (2–)3 mm. in diameter.  Petals ovate-elliptic, glabrous, membranous, whitish or yellowish, (3–)4–5 mm. long, 2–3 mm. broad.  Stamens ± 3 mm. long; filaments glabrous.  Ovary ovoid, laxly pubescent.  Fruit suborbicular, ± 1·7 cm. across, with ± 6 thin transversely nerved, pubescent, distally attenuate and crenate wings ± 1 cm. long.  Seeds ovoid, small.  Fig. 4.

KENYA.  Kwale District: Shimba Hills, Mwele Mdogo forest, Feb. 1953, *Drummond & Hemsley* 1105!
TANZANIA.  Pangani District: Madanga, Nov. 1956, *Tanner* 3320!; Morogoro District: Lusunguru Forest Reserve, Mar. 1953, *Drummond & Hemsley* 1924!; Lindi District: Lake Lutamba, Dec. 1934, *Schlieben* 5715!; Zanzibar I., Haitajwa Hill, Jan. 1929, *Greenway* 1212!
DISTR.  **K**7; **T**3, 6, 8; **Z**; Mozambique
HAB.  Undershrub (sometimes dominant) in lowland evergreen forest, rain-forest, semi-swamp and riverine forest; up to 1525 m.

SYN.  *Oncoba boivinii* (Jaub.) Baill. in Adansonia 10: 250, in text (1872)

## 5. BUCHNERODENDRON

Gürke in E.J. 18: 161, t. 6 (1893); Gilg in E. & P. Pf., ed. 2, 21: 405 (1925);
Sleumer in E.J. 94: 289 (1974)

Subshrubs, shrubs or small trees.  Leaves persistent, alternate, serrate or dentate, penninerved, petiolate; stipules ± caducous.  Flowers in axillary peduncled cymose panicles, racemes or fascicles, the ♂ often subumbellately arranged, bisexual, or unisexual, apparently rarely dioecious, each flower with a rather large subpersistent basal bract.  Sepals 3, valvate, finally free to the base, with setose soft processes on the back.  Petals 6–12, imbricate, larger than the sepals.  Stamens numerous, in 2 series, the outer somewhat longer than the inner one; filaments slender, rather short; anthers linear, dehiscing by slits.  Ovary 1-locular, with 3–5 parietal multi-ovulate placentas; style simple; stigma subentire.  Capsule globose, tardily dehiscent longitudinally into 3–5 valves, covered in numerous elongate branched soft bristles.  Seeds rather numerous, ovoid or compressed; testa crustaceous, arillate at base.

A genus of 2 species in central and eastern Africa, of which 1 is represented in East Africa.

**B. lasiocalyx** (*Oliv.*) *Gilg* in E.J. 40: 467 (1908); V.E. 3 (2): 571 (1921); T.T.C.L.: 230 (1949); Wild in F.Z. 1: 267, t. 44/A (1960); Sleumer in E.J. 94: 293 (1974).  Type: Tanzania, Kilwa, *Kirk* (K, holo.!)

Small shrub or subshrub, (0·1–)0·2–1(–6) m. tall; branches stiffly erect, fuscous tomentose when young, glabrescent and brownish purple with age. Leaf-blades ovate-oblong, broadly ovate or obovate, sometimes obovate-elliptic, apex subacute or obtuse, rarely rounded, base generally cordate,

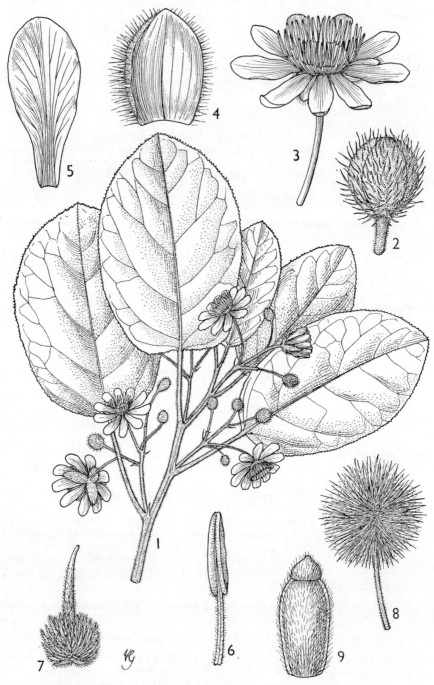

FIG. 5. *BUCHNERODENDRON LASIOCALYX*—**1**, flowering branch, × ⅔; **2**, flower bud, ×4; **3**, flower, × 2; **4**, sepal, × 4; **5**, petal, × 4; **6**, stamen, × 6; **7**, pistil, × 4; **8**, fruit, × ⅔; **9**, seed, × 8. 1, 2, from *Faulkner* 343; 3–7, from *Milne-Redhead & Taylor* 7655; 8, 9, from *Allen* 147. Drawn by Victoria Goaman.

rarely subtruncate-cordate or -rounded, membranous to chartaceous, hairy above, tomentose beneath at least when young, glabrescent with age, serrate to sinuate-dentate, 5–11(–17) cm. long, 3–5(–10) cm. broad; lateral nerves 6–9 pairs (2 basal), slightly prominent above, markedly so beneath, venation reticulate and raised beneath; petiole tomentose, (1–)2–4(–7) cm. long; stipules linear-lanceolate, tomentose, 6–10 mm. long, 2–5 mm. broad at base. Inflorescences in axillary cymose (2–)5–8-flowered panicles, (3–)8–10 cm. long, on a tomentose peduncle (0·5–)1–3 cm. long; pedicels 1–1·3 cm. long; basal bracts narrowly elliptic, hooded, acuminate, serrate, tomentellous, ± 7 mm. long. Flowers bisexual. Sepals ovate-elliptic, acutish, dorsally covered with numerous setose processes, 6–10(–15) mm. long. Petals 8–(12), obovate-oblong, obtuse, white, edge somewhat undulate, dorsally tomentose, 1·5–2 cm. long, 0·5–0·7 cm. broad. Stamens very numerous; filaments puberulous, 3–5 mm.; anthers linear, puberulous, ± 5 mm. Ovary ellipsoid-globose, set with soft setose processes; style slender, ± 5 mm. long. Capsule globose, covered with many puberulous branching bristles or spines, up to 4·5 cm. across including the bristles (1·5–2 cm.), dehiscent into 3 valves. Seeds several, compressed, ± 4 mm. across; testa pubescent; aril dark brown when dry. Fig. 5.

TANZANIA. Rufiji, Jan. 1934, *Musk* 146!; Lindi District: Mahiwa, Dec. 1955, *Milne-Redhead & Taylor* 7655! & Nachingwea, Feb. 1953, *Anderson* 840!
DISTR. T6, 8; Mozambique
HAB. In dense *Brachystegia* woodland or more open bushland; up to 450 m.

SYN. *Oncoba lasiocalyx* Oliv. in Hook., Ic. Pl., t. 1485 (1885)
    *O. eximia* Gilg in E.J. 28: 438 (1900). Type: Tanzania, Kilosa District, E. slope of Vidunda Mts., *Goetze* 409 (B, holo. †, K, iso.!)
    *Buchnerodendron bussei* Gilg in E.J. 40: 468 (1908); V.E. 3 (2): 571 (1921); T.T.C.L.: 230 (1949). Type: Tanzania, Songea District, Mbarangandu R., *Busse* 1288 (B, holo. †, EA, iso.)
    *B. nanum* Gilg in E.J. 40: 467 (1908); V.E. 3 (2): 571 (1921); T.T.C.L.: 230 (1949). Type: Tanzania, Kilwa District, Barikiwa, *Busse* 581 (B, holo. †)
    *B. eximium* (Gilg) Engl., V.E. 1 (1): 390, 406, in text (1910); T.T.C.L.: 230 (1949)

# 6. ONCOBA

Forssk., Fl. Aegypt.-Arab.: CXIII, 103 (1775); Oliv. in F.T.A. 1: 114 (1868); Gilg in E. & P. Pf., ed. 2, 21: 401, fig. 172 (1925)

Shrubs or small trees; stem and branches often spiny. Leaves deciduous, alternate, crenate-serrate, penninerved; petiole not thickened apically; stipules caducous. Flowers bisexual or polygamous (andromonoecious), *Camellia*-like, white, scented, solitary, axillary or terminal on short side shoots, pedunculate. Sepals 3–5, ± united at the base, ± imbricate, the outer ones gradually larger than the inner ones. Petals (5–)8–15, imbricate in bud, subequal. Stamens very numerous; filaments filiform, anthers linear to oblong, basifixed, longitudinally dehiscent, apiculate by the ± elongate connective. Ovary (wanting in ♂ flowers) sessile, unilocular, the 6–8(–10) placentas much protruding into the cavity, multi-ovulate; style columnar; stigma either peltate, i.e. formed by short subglobular and ± connate stigmatic branches (the number of the placentas), or radiate, i.e. the short stigmatic branches free, each with a capitate stigma. Fruit globular, indehiscent, 1-locular, many-seeded, smooth or with slight longitudinal ridges, the pericarp woody. Seeds embedded in an edible pulp, with a smooth bony testa.

A genus of 4 species in tropical (and partly subtropical) Africa, one of them also in and one limited to East Africa.

Leaves rather faintly serrate or crenate-serrate, the
    terminal gland of each tooth rather small, the
    tertiary nerves densely and visibly reticulated
    (fig. 6/2); anthers linear to linear-oblong; stigma
    peltate, with an almost entire or slightly divided
    margin . . . . . . . .    1. *O. spinosa*
Leaves rather coarsely serrate-crenate, the terminal
    gland of each tooth thickish, the tertiary nerves
    parallel to each other, the proper reticulation
    obscure (fig. 6/9); anthers oblong; apex of style
    divided into several distinct branches (3 mm.)
    with capitate stigmas . . . . .    2. *O. routledgei*

1. **O. spinosa** *Forssk.*, Fl. Aegypt.-Arab.: CXIII, 103 (1775); Oliv. in
F.T.A. 1: 115 (1868), incl. var. *angolensis* Oliv.; Engl. in P.O.A. C: 277
(1895); Sim, For. Fl. Port. E. Afr.: 12, t. 2/B (1909); T.S.K.: 21 (1936);
T.T.C.L.: 234 (1949); I.T.U., ed. 2: 149 (1952); Wild in F.Z. 1: 275, t.
46/B (1960); K.T.S.: 227 (1961); F.F.N.R.: 266 (1962); Bamps in F.C.B.,
Flacourt.: 16, fig. 1/A (1968). Type: Yemen, Hadiè & Uadi Surdus,
*Forsskål Herbarium* 626 & 627 (C, syn. !)

Shrub or small tree, up to 10 m. tall, much branched, with bushy crown.
Branches and branchlets with numerous lenticels and straight sharp axillary
spines up to 7 cm. long. Leaf-blades usually elliptic or ovate-elliptic, apex
acuminate, base cuneate, ± papyraceous, rarely subcoriaceous, often wine
red when young, glabrous, margin finely serrate or crenate-serrate (more
coarsely so in coppice shoots), 3·5–14 cm. long, 2–7 cm. broad; lateral
nerves 6–8(–10) pairs, tertiary ones dense and visibly reticulated on both
faces; petiole 0·6–1 cm. long. Flowers solitary, white, beautiful, on a
peduncle 1–2 cm. long. Sepals broadly ovate to suborbicular, greenish
dorsally, white inside, ± persistent and finally reflexed, 1–1·5 cm. long,
0·8–1·2 cm. broad. Petals obovate-cuneate, pure white in anthesis (often
pinkish in bud or when ageing), outer ones 2·5–3·5 cm. long and 1·5–2 cm.
broad, inner ones narrower. Stamens 200–300, ± cohering in 5 bundles;
anthers linear to linear-oblong, 2–3 mm. Ovary subglobose; style 6–10 mm.
long, with a circle of sessile subglobose (± connate) stigma-rays on top.
Fruit round, 5(–6) cm. in diameter, smooth or marked with a number of
slight longitudinal rib-like markings, with a hard shell-like red-brown and
glossy pericarp. Seeds ovoid, blackish brown, glabrous, 6–7 mm. long,
3–4 mm. broad. Fig. 6/1–8.

UGANDA. W. Nile District: Moyo Station, Sept. 1937, *Eggeling* 3423!; Kigezi
    District: Bwambara, Feb. 1950, *Purseglove* 3269!; Mbale District: Bugisu, Bududa,
    July 1926, *Maitland* 1227!
KENYA. Northern Frontier Province: Moyale, July 1952, *Gillett* 13504!; Turkana
    District: Kiriamet, Mar. 1961, *Tweedie* 2115!; Lamu District: Kitangani, Dec. 1946,
    *J. Adamson* 314!
TANZANIA. Mwanza District: Ukerewe I., Mar. 1928, *Conrads* 492!; Moshi District:
    Kware R., Dec. 1893, *Volkens* 1606!; Singida District: Wamba [Iwumbu] R., Aug.
    1927, *B. D. Burtt* 730!; Zanzibar I., without locality, *Toms* 107!
DISTR. U1–4; K1–5, 7; T1, 2, 4–7; Z; widespread in tropical Africa, also in Arabia and
    South Africa (Transvaal)
HAB. Forest edges, riverine forest and bushland, *Brachystegia* woodland; up to
    1800 m.

NOTE. Sometimes grown around Nairobi as an ornamental.

2. **O. routledgei** *Sprague* in Gard. Chron., ser. 3, 49: 323, figs. 145 & 146
(1911) & in K.B. 1911 : 262 (1911); R.E. Fries in N.B.G.B. 9: 323 (1925);

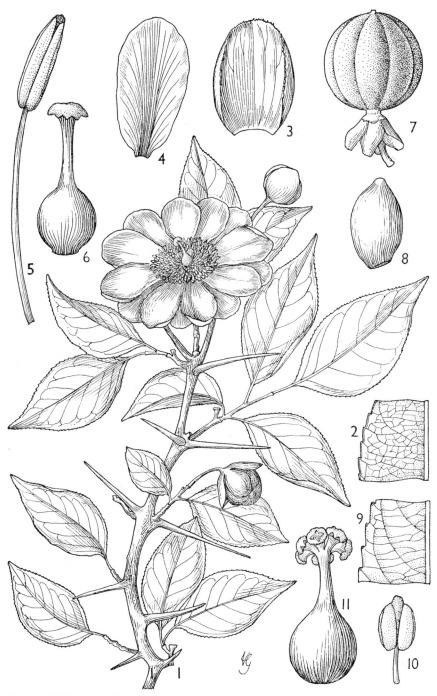

Fig. 6. *ONCOBA SPINOSA*—**1**, flowering branch, × ⅔; **2**, detail of leaf margin and venation, × 2; **3**, sepal, × 2; **4**, petal, × 1; **5**, stamen, × 10; **6**, pistil, × 3; **7**, fruit, × ⅔; **8**, seed, × 5. *O. ROUTLEDGEI*—**9**, detail of leaf, × 2; **10**, stamen, × 10; **11**, pistil, × 3. 1–4, from *Greenway & Kanuri* 12034; 5, 6, from *Richards* 18278; 7, 8, from *Gillett* 13504; 9–11, from *Gibson* 388. Drawn by Victoria Goaman.

T.S.K.: 21 (1936); T.T.C.L.: 234 (1949); I.T.U., ed. 2: 149 (1952); Bamps in F.C.B., Flacourt.: 18, fig. 1/B (1968). Type: Uganda, Ruwenzori, 2400 m., *Dawe* 650 (K, holo.!)

Shrub or small tree, up to 8 m. tall; branches with straight spines up to 2·5 cm. long. Leaf-blades elliptic-ovate or -oblong, apex shortly acuminate, base obtuse to rounded, papyraceous to subcoriaceous, glabrous, dark green and glossy, rather coarsely serrate-crenate, 6–16 cm. long, 3–6 cm. broad; lateral nerves 4–6 pairs, tertiary ones ± parallel, rather obvious on both faces, the proper reticulation being ± obscure; petiole 0·4–1 cm. long. Flowers solitary, white and somewhat waxy, much resembling those of *O. spinosa*, on a peduncle 1–2 cm. long. Sepals suborbicular, ± 1 cm. in diameter. Petals obovate-oblong, ± 2·5 cm. long and 1·5 cm. broad. Stamens with oblong anthers (1·5–2 mm.). Style 5 mm. long, the 6–8 free stigmatic and ± recurved branches ± 3 mm. long, with capitate stigmas. Fruit subglobular, 4–4·5 cm. long, 5·5–6 cm. across, yellow, smooth, with 8 longitudinal slightly prominent markings. Seeds ovoid, ± flattened, 6–7 × 3–4 mm., light brown. Fig. 6/9–11.

UGANDA. Ankole District: Kalinzu Forest, June 1938, *Eggeling* 3731!; Kigezi District: Kayonza, Marambo, Mar. 1947, *Purseglove* 2383!; Mbale District: Bugisu, Butandiga, Dec. 1938, *A. S. Thomas* 2566!
KENYA. Laikipia plateau, 1908, *Routledge*!; Kiambu District: Limuru, June 1918, *Snowden* 586!; Kericho District: Sambret Catchment, June 1958, *Kerfoot* 169!
TANZANIA. Arusha District: Meru Forest Reserve, Nov. 1955, *Sangiwa* 109!; Lushoto District: Mt. Gonja, July 1893, *Holst* 4217!; Morogoro District: Uluguru Mts., Bunduki, Jan. 1935, *E. M. Bruce* 439!
DISTR. U2, 3; K?3, 4, 5; T2, 3, 6; also in Ethiopia and Zaire (Lakes Albert, Edward and Kivu)
HAB. Rain-forest, often riparian; 900–2440 m.

SYN. *O. micrantha* Gilg in Engl., V.E. 3 (2): 565, in obs. (1921) & in E. & P. Pf., ed. 2, 21: 402 (1925). Type: Zaire, between Beni and Irumu R., *Mildbraed* 2810 (B, holo. †)
    *O. spinosa* Forssk. var. *routledgei* (Sprague) Dale & Greenway, K.T.S.: 227 (1961), comb. non rite publ.


## 7. XYLOTHECA

### Hochst. in Flora 26: 69 (1843)

Shrubs or small trees, unarmed. Leaves deciduous, alternate, thin to leathery at full maturity, entire or undulate, petiolate, penninerved; stipules small, caducous. Flowers usually large and showy, white, sweet-scented, bisexual or andromonoecious, cymose or umbellate in the upper leaf-axils, or terminal on short branchlets. Sepals 3 (or 4), concave, free or almost so, imbricate, rather early caducous, glabrous or pubescent, often with sessile resinous glands. Petals 7–14, free, narrowed to the base, imbricate, spreading at anthesis. Stamens 20–30; filaments free, filiform; anthers linear, basi-fixed, dehiscing longitudinally from above. Ovary (rudimentary or absent in the ♂ flowers) sessile, 1-locular, the 6 or 7 placentas multi-ovulate; style terminal, columnar; stigmas as many as placentas, rather short and spreading. Fruit a woody capsule splitting into (4–)6–8 longitudinal rather thick valves; style persistent as a hard apical point. Seeds numerous, ellipsoid, arillate or embedded in a thin pulp.

A genus of 2 species in SE. and E. Africa, and 1 in S. Angola, with 1 species (and 3 varieties) in East Africa.

*X. kraussiana* Hochst., from southern Africa, has been grown around Nairobi, e.g. *Grahame Bell* 3.

**X. tettensis** (*Klotzsch*) *Gilg* in E.J. 40: 456 (1908); V.E. 3(2): 567 (1921); T.T.C.L.: 237 (1949); Wild in Bol. Soc. Brot., sér. 2, 32: 53 (1958) & in F.Z. 1: 274 (1960). Type: Mozambique, Sena, *Peters* (B, holo. †)

Shrub or small tree, 1–2(–5) m. tall; young branches densely yellowish hairy or glabrous. Leaves appearing just before the flowers; blade obovate or obovate-oblong, apex rounded, base rounded or subcordate, chartaceous in the adult stage, glabrous to densely pubescent on both faces, entire or undulate, (1·5–)2–9(–11) cm. long, 0·6–6(–8) cm. broad; lateral nerves 4–8 pairs, looping within the edge, somewhat prominent beneath as is the reticulation; petiole up to 1·5(–3) cm. long. Flowers usually solitary, occasionally in 2–3-flowered cymes in the leaf-axils or terminal on (abbreviated) branchlets; peduncle short; pedicels up to 2 cm. long; basal bracts deltoid, caducous, ± 2 mm. long. Sepals (3 or) 4, oblong-suborbicular, the margin membranous, glabrous to pubescent and with resinous glands outside, 1–2 cm. long, (0·6–)1–1·5(–2) cm. broad. Petals 7–12, narrowly obovate, cuneate to a shortly clawed base, glabrous, up to 4·5 cm. long and 2·4 cm. broad. Stamens numerous; filaments glabrous, up to 1 cm. long; anthers linear, (3–)4–6 mm. long. Ovary ovoid, densely hairy, smooth or longitudinally sulcate; style columnar, pilose below, up to 1 cm. long; stigmas recurved, ± 2 mm. long. Fruit woody, subglobose to ovoid, yellowish hairy to glabrous, smooth or sulcate, ± 3·5–4 cm. long and 1·7–2·5 cm. across, splitting into ± 8 longitudinal segments. Seeds many, ellipsoid, pale brown or buff, smooth or rugose, glabrous, not obviously arillate but embedded in a thin edible pulp.

### KEY TO INFRASPECIFIC VARIANTS

Vegetative parts sparsely to densely pubescent:
  Leaf-blades usually less than 3 cm. long; fruit
    usually rather deeply 12–14-sulcate . . a. var. **tettensis**
  Leaf-blades usually exceeding 3 cm. in length;
    fruit smooth . . . . . . b. var. **macrophylla**
Vegetative parts glabrous or practically so (the young
  ones sometimes noticeably glutinous with resinous
  glands):
  Fruit smooth or slightly sulcate . . . . c. var. **kirkii**
  Fruit ± deeply 10-sulcate, the ribs crenately
    incised . . . . . . . d. var. **fissistyla**

a. var. **tettensis**; Wild in Bol. Soc. Brot., sér. 2, 32: 54 (1958) & in F.Z. 1: 274 (1960)

Young branches usually densely yellowish hairy, sometimes glabrous. Leaf-blades (1·5–)2–3 cm. long, 0·6–1·6 cm. broad, sparsely pubescent; lateral nerves 4–6 pairs. Ovary and fruit 12–14-sulcate.

TANZANIA. Masasi District: 19 km. W. of Mahiwa, Dec. 1955, *Milne-Redhead & Taylor* 7662!; Lindi District: Mchinjiri, Rondo Forest Reserve, Nov. 1952, *Wigg* 1040! & Lake Lutamba, Dec. 1934, *Schlieben* 5705!
DISTR. **T**6, 8; Malawi, Mozambique
HAB. Scattered in lowland woodland or secondary bushland, on red sandy soil; ± 200–600 m.

SYN. *Chlanis tettensis* Klotzsch in Peters, Reise Mossamb., Bot. 1: 145 (1861)
  *Oncoba tettensis* (Klotzsch) Harv. in Fl. Cap. 2: 584 (1862), pro parte; Oliv. in F.T.A. 1: 116 (1868)

b. var. **macrophylla** (*Klotzsch*) *Wild* in Bol. Soc. Brot., sér. 2, 32: 54 (1958) & in F.Z. 1: 275, t. 46/A (1960). Type: Mozambique, Sena, *Peters* (B, holo. †)

Branches and branchlets densely long-hairy. Mature leaf-blades (5–)7–9(–11) cm. long, 4–6(–7·5) cm. broad, ± densely pubescent; lateral nerves 6–8 pairs. Fruit usually smooth.

FIG. 7. *XYLOTHECA TETTENSIS* var. *KIRKII*—**1,** flowering branch, × ⅔; **2,** sepal, × 2; **3,** petal, × 2; **4,** stamen, × 8; **5,** pistil, × 4; **6,** fruit, × 1; **7,** fruit dehiscing, × 1; **8,** seed, × 5. 1–4, from *Faulkner* 1760; 5, from *Milne-Redhead & Taylor* 7353; 6, 8, from *Drummond & Hemsley* 1823; 7, from *Thulin & Mhoro* 476. Drawn by Victoria Goaman.

TANZANIA. Rufiji, Dec. 1930, *Musk* 155!; Lindi District: S. of R. Mbemkuru, Dec. 1955, *Milne-Redhead & Taylor* 7579!; Mikindani District: Mtwara–Mikindani road, Mar. 1963, *Richards* 17865!

DISTR. T6, 8; Malawi, Mozambique

HAB. Lowland woodland or bushland, as var. *tettensis*; 0–600 m.

SYN. *Chlanis macrophylla* Klotzsch in Peters, Reise Mossamb., Bot. 1: 145 (1861)
   *Oncoba petersiana* Oliv. in F.T.A. 1: 116 (1868), *nom. illegit.* (based on *Chlanis macrophylla*)
   *O. macrophylla* (Klotzsch) Warb. in E. & P. Pf. III. 6a: 18 (1893) & in P.O.A. C: 277 (1895)
   *O. stuhlmannii* Gürke in E.J. 18: 164 (1893). Type: Mozambique, Quelimane, *Stuhlmann* ser. 1, 707 (B, holo. †, HBG, iso.!)
   *O. angustipetala* De Wild., Pl. Nov. Herb. Hort. Then. 1: 13, t. 4 (1904). Type: Mozambique, Morrumbala, *Luja* 395 (BR, holo.!)
   *Xylotheca stuhlmannii* (Gürke) Gilg in E.J. 40: 456 (1908)
   *X. macrophylla* (Klotzsch) Sleumer in F.R. 45: 20 (1938); T.T.C.L.: 237 (1949)

NOTE. The differentiation from var. *tettensis* is possibly rather arbitrary.

c. var. **kirkii** (*Oliv.*) *Wild* in Bol. Soc. Brot., sér. 2, 32: 55 (1958) & in F.Z. 1: 275 (1960); K.T.S.: 229 (1961). Type: Tanzania/Mozambique, Rovuma Bay, *Kirk* (K, holo.!)

Leaf-blades 6–8(–11) cm. long, 2·5–4·5(–8) cm. broad, i.e. like those of var. *macrophylla*, but the vegetative parts are glabrous. The young parts are often noticeably glutinous with resin glands. Ovary and fruit smooth or slightly 12–14-sulcate. Fig. 7.

KENYA. Kilifi District: Ganda, Feb. 1938, *Dale* in *F.D.* 3879!

TANZANIA. Tanga District: Pongwe, Jan. 1937, *Greenway* 4839!; Morogoro District: Mtibwa, Mar. 1953, *Drummond & Hemsley* 1823!; Rufiji District: Mafia I., Mito Miwili, Aug. 1937, *Greenway* 5030!; Lindi District: Mchinjiri, Sept. 1951, *Bryce* B. 13!; Zanzibar I., Feb. 1874, *Hildebrandt* 1175!

DISTR. K7; T3, 6, 8; Z; Mozambique

HAB. Lowland woodland, bushland and wooded grassland similar to that of var. *tettensis* and var. *macrophylla*; 0–700 m.

SYN. *Oncoba kirkii* Oliv. in F.T.A. 1: 116 (1868)
   *Xylotheca kirkii* (Oliv.) Gilg in E.J. 40: 455 (1908); T.T.C.L.: 236 (1949)
   *X. glutinosa* Gilg in E.J. 40: 457 (1908); T.T.C.L.: 236 (1949). Type: Tanzania, Uluguru Mts., Tunanguo, *Stuhlmann* 8979 (B, holo. †)

var. **fissistyla** (*Warb.*) *Sleumer*, stat. nov. Type: Tanzania, Bagamoyo, *Stuhlmann* 125 (B, holo. †, K, iso.!)

Leaf-blades 5–9(–11) cm. long, 2–5(–6) cm. broad, glabrous, i.e. like those of var. *kirkii*. Ovary pubescent, deeply 10-sulcate; stigmata 7, ± 3 mm. Fruit narrowly ovoid, ± 4 cm. long, 1·7–2 cm. across, pubescent, deeply 10-sulcate, the ribs distinctly crenately incised.

TANZANIA. Bagamoyo, Dec. 1915, *Peter* 14735a!; Uzaramo District: Dar es Salaam, 16 km. along Bagamoyo road, Apr. 1933, *B. D. Burtt* 4467! & Pugu Hills near Kisarawe, Oct. 1968, *Harris & Walker* 2504!

DISTR. T6; not known elsewhere

HAB. Apparently limited to lowland woodland, coastal bushland and wooded grassland, and possibly only a form of var. *kirkii*; 0–200 m.

SYN. *Oncoba fissistyla* Warb. in P.O.A. C: 277 (1895)
   *Xylotheca fissistyla* (Warb.) Gilg in E.J. 40: 455 (1908); T.T.C.L.: 236 (1949)
   *X. holtzii* Gilg in E.J. 40: 456 (1908); T.T.C.L.: 236 (1949). Type: Tanzania, Uzaramo District, Mogo Forest [Sachsenwald], *Holtz* 380 (B, holo. †)
   *X. sulcata* Gilg in E.J. 40: 456 (1908); T.T.C.L.: 237 (1949). Type: Tanzania, Uzaramo District, Vikindu, *Stuhlmann* 6137 (B, holo. †)

# 8. CALONCOBA

Gilg in E.J. 40: 458 (1908) & in E. & P. Pf., ed. 2, 21: 402 (1925); Sleumer in E.J. 94: 120 (1974)

Shrubs or trees. Leaves persistent or deciduous, alternate, entire to undulate, petiolate, glabrous or glandular-punctate, quite often glutinous when young; stipules ± caducous. Flowers rather large, in abbreviated racemes, or in fascicles, or solitary, axillary or from old branches or from the trunk, bisexual and ♂ in the same specimen, sometimes appearing before the leaves. Sepals 3, imbricate, concave. Petals 8–13, ± twice as large as the sepals, obovate or oblanceolate, ± clawed, thin, white, scented. Stamens very numerous; filaments filiform; anthers linear-subsagittate, dehiscing by apical pores or ± elongate longitudinal slits. Ovary 1-locular, with 5–8 multi-ovulate placentas; style simple, distally truncate-subcapitate or divided into 5–8 short stigmatic lobes or branches. Fruit an echinate or smooth, dehiscent ovoid-subglobose to slightly obovoid capsule splitting into 5–8 valves, many-seeded, the seeds generally embedded in a fleshy or gelatinous pulp.

A genus of 10 species in Africa, of which 2 are represented in East Africa.

Ovary echinulate; fruit with slender almost woody
  spines    .    .    .    .    .    .    .    .    1. *C. welwitschii*
Ovary and fruit smooth    .    .    .    .    .    .    2. *C. crepiniana*

1. **C. welwitschii** (*Oliv.*) *Gilg* in E.J. 40: 462 (1908); V.E. 3 (2): 568 (1921); T.T.C.L.: 230 (1949); Sleumer in E.J. 94: 124 (1974). Type: Angola, Cuanza Norte, Golungo Alto, *Welwitsch* 537 (LISU, holo., BM, BR, G, K, P, iso.!)

Shrub or generally tree, 4–14(–20) m. tall; bark greyish-brownish. Leaves ± deciduous, aggregated towards the ends of the branches; blade ovate, apex acuminate, base broadly cuneate to cordate, membranous to chartaceous, glabrous, 10–28 cm. long, (5·5–)8–20 cm. broad, 5-nerved from the base, upper lateral nerves 2–4 pairs, veins ± transverse, prominent beneath with ± pronounced reticulation of the veinlets; petiole 4–12(–19) cm. long; stipules subulate-aristate, up to 2·5 cm. long, caducous. Flowers 2–5 in a fascicle from defoliate axils, or from old branches, even from the trunk, appearing with the young leaves, up to 10 cm. in diameter; pedicels sparingly glandular, 1·5–2·5 cm. long at anthesis, up to 4 cm. in fruit. Sepals oblong-elliptic, glandular dorsally, subpersistent, 1·2–2 cm. long, 0·7–1·3 cm. broad. Petals ± 10, about twice the size of the sepals, spathulate-oblong, strongly veined towards the base, (1·6–)2·5–4·5 cm. long, (0·6–)1–2·3 cm. broad. Filaments up to 2·5 cm. long; anthers 3–4 mm. long. Ovary tuberculate; style slender, ± 1 cm. long, with 5–6 short stigmatic branches. Capsule elliptic to subglobular, 7–10(–15) cm. long, 4–6(–12) cm. across including the spines (these (1·5–)2–5 cm. long), splitting into 5–6 valves, which recurve when ripe; pericarp ± 5 mm. thick. Seeds numerous, globose, puberulous, 5–7 mm. long, 3–4 mm. broad. Fig. 8/8, 9.

TANZANIA. Lushoto District: E. Usambara Mts., Lutindi Peak, Jan. 1941, *Greenway* 6136!; Njombe District: Lupembe, Ruhudji R., Oct. 1931, *Schlieben* 1346!; Lindi District: Rondo Plateau, Mchinjiri, Dec. 1955, *Milne-Redhead & Taylor* 7602!
DISTR. **T**3, 6–8; Nigeria to Zaire and Angola, also in Malawi and Mozambique
HAB. In the lower storey of rain-forest, dry evergreen forest and riverine forest, also in secondary growth; 800–1900 m.

SYN. *Oncoba welwitschii* Oliv., F.T.A. 1: 117 (1868)
    *Caloncoba gigantocarpa* Perkins & Gilg in E.J. 40: 464 (1908); T.T.C.L.: 230 (1949). Type: Tanzania, Newala/Masasi Districts, Makonde Plateau, Mkomadatchi, *Busse* 1092 (B, holo. †, EA, iso.!)
    *C. grotei* Gilg in Engl., V.E. 3 (2): 568 (1921), in text; T.T.C.L.: 230 (1949). Type: Tanzania, cultivated at Amani, *Herb. Amani* 5643 (B, holo. †, EA, iso.!)
    *C. cauliflora* Sleumer in N.B.G.B. 12: 86 (1934). Type: Tanzania, Uluguru Mts., 1200 m., *Schlieben* 4254 (B, holo. †, B, BM, BR, M, P, iso.!)

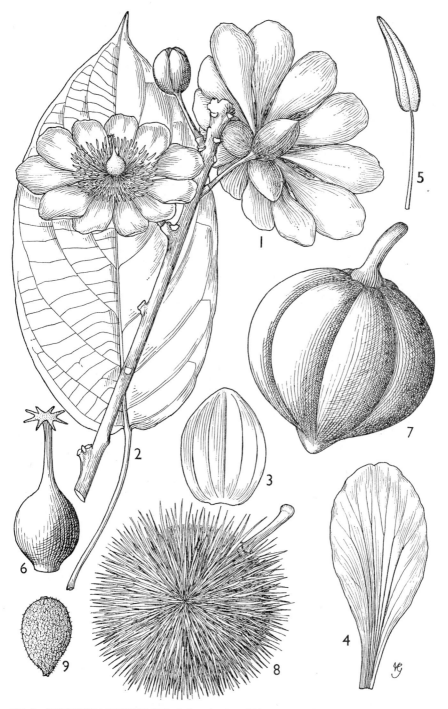

FIG. 8. *CALONCOBA CREPINIANA*—**1**, flowering branchlet, × ⅔; **2**, leaf, × ⅔; **3**, sepal, × 1⅓; **4**, petal, × 1⅓; **5**, stamen, × 6; **6**, pistil, × 2; **7**, fruit, × ⅔. *C. WELWITSCHII*—**8**, fruit, × ⅔; **9**, seed, × 3. 1, 2, from *Harris* 458; 3–6, from *Schweinfurth* 2964; 7, from *Sillitoe* 492; 8, 9, from *Milne-Redhead & Taylor* 7602. Drawn by Victoria Goaman.

2. **C. crepiniana** (*De Wild. & Th. Dur.*) *Gilg* in E.J. 40: 460 (1908); V.E. 3 (2): 568 (1921); Bamps in F.C.B., Flacourt.: 24 (1968); Sleumer in E.J. 94: 136 (1974). Type: Zaire, Bikoro, Lukolela, *Dewèvre* 848 (BR, holo.!)

Shrub or tree (rarely up to 25 m. tall); bark pale grey to brownish. Leaves deciduous; blade ovate or elliptic, sometimes almost obovate, apex obtusely acuminate, base broadly cuneate to rounded, or even subcordate, juvenile ones papyraceous, subcoriaceous with age, initially microglandular and glutinose, finally glabrous, entire, 8–26 cm. long, 4–15 cm. broad; lateral nerves 5–7 pairs prominent beneath, reticulation lax, a little raised beneath; petiole (2·5–)4–10 cm. long; stipules linear, caducous, 2–5 mm. long. Flowers solitary or 2 or 3 in a fascicle from leafless axils of the one year old shoots, 7–10 cm. across; pedicels 2–4 cm. long at anthesis, accrescent to 5 cm. in fruit. Sepals ovate-suborbicular, glandular dorsally, persistent during anthesis, 1·6–2·5 cm. long, 1·1–1·7 cm. broad. Petals 12 or 13, obovate to oblanceolate, shortly clawed, 3·5–5·5 cm. long, 1·2–2·7 cm. broad. Filaments 1·5–2·5 cm. long; anthers mucronulate, 4–5 mm. long. Ovary covered with warty glands; style ± 1·2 cm. long, ending in 6–8 stigmatic branches ± 3 mm. long. Capsule subglobular, broadly conical or beaked distally, at first green mottled with white, turning orange as it ripens, (5–)7(–8) cm. long, 5–6(–8) cm. across; pericarp 8–10 mm. thick. Seeds numerous, angular, glabrous, white, 7–8 mm. long, 5–6 mm. broad. Fig. 8/1–7.

UGANDA. W. Nile District: Leya R. headwaters, Mar. 1945, *Greenway & Eggeling* 7257!; Bunyoro District: Budongo Forest, Jan. 1963, *Styles* 327!; Toro District: Bwamba, Mar. 1939, *A. S. Thomas* 2803!
DISTR. U1, 2, (*fide* I.T.U.) 4; Central African Republic, S. Sudan, Zaire
HAB. Understorey and margins of rain-forest and riverine forest, also in secondary growth and wooded grassland; 850–1500 m.

SYN. *Oncoba crepiniana* De Wild. & Th. Dur., Ann. Mus. Congo, Bot., sér. 2, 1: 7 (1899)
     *Caloncoba schweinfurthii* Gilg in E.J. 40: 461 (1908); I.T.U., ed. 2: 144, fig. 31 (1952). Type: Zaire, Ubangi-Uele, Niamniam, *Schweinfurth* 2964 (B, lecto. †, BR, K, P, WU, isolecto.!)

## 9. LINDACKERIA

Presl, Rel. Haenk. 2: 89, t. 65 (1835); Sleumer in E.J. 94: 311 (1974)

Shrubs or trees. Leaves persistent or deciduous, alternate, subentire or generally dentate, penninerved, ± long-petiolate, stipulate. Flowers bisexual or ♂ by abortion, in axillary racemes or panicles, rarely reduced to a solitary flower. Sepals (2 or) 3, imbricate, concave, caducous. Petals 6–12, imbricate, slightly longer than the sepals, white, caducous. Stamens indefinite; filaments slender, free or rarely conglutinate below; anthers linear, opening lengthwise. Ovary sometimes shortly stalked, smooth, tuberculate-muriculate or shortly echinate, 1-locular, with 3 (or 4) few- to multi-ovulate placentas; style filiform, entire or rarely shortly divided. Capsule woody, with short obtuse emergences or with spines, tardily dehiscent into 3 (or 4) valves. Seeds 1–3, rarely more, mutually compressed, partly covered with a red aril, with copious endosperm.

A genus with about 6 tropical American, and 7 tropical African species, 4 of them in East Africa.

Petals 2–2·3 cm. long; capsule with glabrous spines;
    flowers solitary or in twos  .  .  .  .  1. *L. fragrans*
Petals up to 1·4 cm. long; capsule with hairy
    emergences or spines:

Pedicels articulate at base; seeds glabrous; leaves
　　(at least the young ones) ± densely pubescent　2. *L. bukobensis*
Pedicels articulate well above the base or in the
　　middle; seeds (laxly) hairy:
　　Leaves glabrous except sometimes on the nerves;
　　　fruit with 3–6(–8) mm. long spines when
　　　fully developed　.　.　.　.　.　3. *L. schweinfurthii*
　　Leaves densely pubescent (or rather velutinous
　　　when young); fruit with 1 mm. long conical
　　　obtuse emergences when fully developed　.　4. *L. stipulata*

1. **L. fragrans** (*Gilg*) *Gilg* in E.J. 40: 466 (1908); R.E. Fries, Wiss.
Ergebn. Schwed. Rhod.-Kongo-Exped. 1: 155 (1914); V.E. 3 (2): 569
(1921); T.T.C.L.: 233 (1949); Wild in F.Z. 1: 269 (1960); F.F.N.R.: 266
(1962); Sleumer in E.J. 94: 315 (1974). Type: Chunya/Mbeya Districts,
Unyiha, Toola, *Goetze* 1411 (B, holo. †, BR, Z, iso.!)

Shrub or small tree up to 6 m. tall. Young branchlets silky-pubescent,
finally glabrescent and densely lenticelled. Leaves ± deciduous, obovate-
or elliptic-oblong, apex shortly and rather abruptly acuminate, base cuneate
to rounded, membranaceous to thin chartaceous, finely and regularly
serrate to serrulate, silky to subpatently hairy on both sides especially on
the midrib when young, glabrescent except the midrib above, and midrib
and nerves beneath with age, (4–)6–9(–16) cm. long, (1·5–)2·5–6 cm. broad;
lateral nerves 8–9 pairs, raised beneath only, reticulation rather obscure;
petioles 1–1·3 cm. long; stipules lanceolate, ± 5 mm. long, 1–2 mm. broad,
caducous. Flowers solitary (or rarely in twos) in the upper axils. Pedicels
subdensely and ± patently hairy, ± 2 cm. long at anthesis, accrescent to
3·5 cm. in fruit, articulate in the lower ¼–⅓. Sepals broadly elliptic, edge
membranous, silky-pubescent dorsally, (8–)10–12 mm. long, 5–6 mm. broad.
Petals 6–8, oblong-obovate, clawed, 2–2·3 cm. long, (6–)8(–13) mm. broad.
Filaments 3–5 mm. long; anthers glabrous, 2·5 mm. Ovary muriculate,
glabrous; style 6–8 mm. long. Capsule not known in fully mature state;
spines of immature fruits inflated at the base, glabrous, 5–6 mm. long.

TANZANIA. Ufipa District: Chala Mt., 13 Dec. 1956, *Richards* 7247!; Chunya/Mbeya
　　District: Unyiha, Toola, Nov. 1899, *Goetze* 1411!
DISTR. T4, 7; Zambia
HAB. By streams in deciduous woodland and riverine forest; 1300–2100 m.

SYN. *Oncoba fragrans* Gilg in E.J. 30: 357 (1901)

2. **L. bukobensis** *Gilg* in E.J. 40: 465 (1908); V.E. 3(2): 569 (1921);
T.T.C.L.: 233 (1949); Wild in F.Z. 1: 269, t. 44/B (1960); F.F.N.R.: 266
(1962); Bamps in F.C.B., Flacourt.: 34 (1968); Sleumer in E.J. 94: 320
(1974). Type: Tanzania, Bukoba, *Stuhlmann* 3895 (B, lecto. †, BR, K,
isolecto.!)

Shrub or tree, rarely up to 12 m. tall; bark greyish-blackish, thin, rough.
Branchlets brownish pubescent to subtomentose at tips, older parts
glabrescent and reddish brown, lenticellate. Leaves ± deciduous; blade
variable in shape, oblanceolate to oblong or elliptic, sometimes obovate-
oblong or even obovate, apex shortly acuminate, subacute or obtuse, base
attenuate to obtuse, rarely rounded or subcordate, chartaceous, initially
± densely hairy to subtomentose on both faces, glabrescent with age,
midrib and nerves of the undersurface remaining ± densely hairy, subentire to rather coarsely sinuate-serrate or -dentate, (7–)9–20(–27) cm. long,
(2·5–)4–9(–12) cm. broad; lateral nerves 6–8(–10) pairs, prominent beneath,

FIG. 9.  *LINDACKERIA BUKOBENSIS*—**1,** flowering branch, × ⅔; **2,** flower, × 3; **3,** sepal, × 6; **4,** petal, × 6; **5,** stamen, × 8; **6,** pistil, × 8; **7,** fruit, × 2; **8,** seed, × 3.  1, 3–6, from *E. Brown* 111; 2, from *Purseglove* 2341; 7, 8, from *Maitland* 802.  Drawn by Victoria Goaman.

veins laxly reticulate, slightly raised beneath; petiole pubescent, 1·5–4(–7) cm.
long; stipules linear-lanceolate, pubescent, 5–8(–10) mm. long, (0·5–)1–2 mm.
broad at base, caducous, sometimes tardily so. Flowers 3–7 in slender,
sometimes subumbelliform, racemes or panicles, yellowish tomentellous;
peduncle (1–)1·5–3(–5) cm. long; rhachis 0·5–1·5 cm. long, sometimes almost
0; pedicels articulate at the base, 1·5–2·5(–3·5) cm. long; bracts lanceolate,
caducous, 2–4 mm. long, 0·5–2 mm. broad at base. Sepals obovate,
tomentellous outside, glabrous inside, (6–)7–10 mm. long, 4–6(–7) mm. broad.
Petals 8(–10), obovate, clawed, hairy all over or at least at the base outside,
and at the base only inside, 9–12(–13) mm. long, (4–)5–7 mm. broad.
Filaments puberulous, 2·5–4 mm. long; anthers pubescent, (2–)2·5–3 mm.
long. Ovary tomentose and set with short bristles; style simple, hairy
except towards the narrow apex. Capsule subglobose, pale apricot colour,
(1·3–)2–2·5(–3) cm. in diameter with the spines included, the latter
pubescent, subconical-swollen at the base, ± elongate-attenuate to
(3–)5–10(–15) mm. distally in fully mature (may be shorter in empty)
fruits. Seeds 1–3(–4), subglabrous, blackish-brownish, (5–)6–8 mm. long,
4–5 mm. broad. Fig. 9.

UGANDA. Bunyoro District: Budongo Forest, Apr. 1938, *Eggeling* 3591!; Kigezi
District: Kayonza, Apr. 1948, *Purseglove* 2640!; Masaka District: Bunyama I.,
June 1925, *Maitland* 802!
KENYA. Lamu District: Kiunga, Dec. 1946, *J. Adamson* 271!
TANZANIA. Kigoma District: Tubila, Nov. 1956, *Procter* 578!; Mpanda District:
Kungwe-Mahali Peninsula, Musenabantu, Aug. 1959, *Harley* 9359!; Rungwe District:
Kyimbila, Jan. 1913, *Stolz* 1839!
DISTR. U2, 4; K7; T1, 4, 6, 7, 8; SE. Somali Republic, Sudan (Equatoria), Zaire,
Rwanda, Burundi, Zambia, Malawi
HAB. Forest or forest edge, often in open mixed forest; (50–)900–1525(–1750) m.

SYN. [*Oncoba dentata* sensu Warb. in P.O.A. C: 277 (1895), *non* Oliv. Based on
Stuhlmann's collection from Bukoba]
*Lindackeria mildbraedii* Gilg in Z.A.E.: 567 (1913); I.T.U., ed. 2: 149 (1952);
Bamps in F.C.B., Flacourt.: 36 (1968). Type: Zaire, Lake Kivu, Idjwi I.,
*Mildbraed* 1208 (B, holo. †, B, iso.!)
*L. somalensis* Chiov. in Result. Sci. Miss. Stef.-Paoli, Coll. Bot.: 24 (1916);
E.P.A.: 596 (1959). Type: SE. Somali Republic, Hididle–Hamagudu, *Paoli*
676 (FI, holo.!)
*L. bequaertii* De Wild., Pl. Bequaert. 5: 407 (1932); Bamps in F.C.B., Flacourt.:
33 (1968). Type: Zaire, Djugu, Kilo, *Bequaert* 4864 (BR, holo.!)
[*L. dentata* sensu T.T.C.L.: 233 (1949), *non* (Oliv.) Gilg]
[*L. sp. aff. grewioides* Sleumer sensu K.T.S.: 226 (1961), based on *J. Adamson*
271 from Kenya, Lamu District]
*L. kivuensis* Bamps in B.J.B.B. 34: 497 (1964) & in F.C.B., Flacourt.: 33 (1968).
Type: Zaire, Bulembi, Nyabiondo, *Gutzwiller* 3253 (BR, holo.!)

NOTE. *L. bukobensis* is conceived here in a broad sense. Part of the (mainly western)
specimens have leaves narrowed at both ends (the " *bequaertii* " and " *mildbraedii* "
type), but cannot be separated from other specimens with broader, ± obovate leaves
(" *bukobensis* ", " *kivuensis* ", and " *somalensis* " type), as intermediates occur, and
no distinctive characters can be found in flowers and fruits. The Kenya specimen
has flowers only, and its determination is not beyond question.

3. **L. schweinfurthii** *Gilg* in E.J. 40: 466 (1908), pro parte; V.E. 3 (2):
569 (1921); F.P.S. 1: 157 (1950); Bamps in F.C.B., Flacourt.: 30, t. 4
(1968); Sleumer in E.J. 94: 322 (1974). Types: Sudan, Equatoria, Niam-
Niam, Nduku [Linduku] R., *Schweinfurth* 3074 (B, lecto. †, K, isolecto.!);
Zaire, Lac Édouard, Issango, *Stuhlmann* 2525 (B, syn. †)

Shrub or tree up to 7 m. tall. Branchlets subglabrous. Leaves persistent;
blade lanceolate-oblong, elliptic or oblanceolate, long acutely acuminate,
base attenuate, membranaceous to papyraceous, generally glabrous on both
faces (sometimes sparsely and shortly hairy along midrib and nerves beneath

only), subentire to crenate-dentate or -subserrate, (8–)12–26 cm. long,
(2·5–)3–6·5(–9) cm. broad; midrib distinctly prominent on both faces;
lateral nerves (6–)7–10 ± looping pairs, slightly raised as is the rather lax
reticulation beneath; petiole 0·5–1·5(–2·5) cm. long; stipules linear,
pubescent, ± 5 mm. long and 0·5 mm. broad, tardily caducous. Racemes
(2–)3–6-flowered, slender, shortly pubescent, on a peduncle (or rhachis)
1–1·5(–2) cm. long; pedicels slender, at anthesis 1–2, in fruit 2–2·5 cm.
long, articulate 3–8 mm. above the base. Sepals subovate-oblong,
pubescent outside, ± 6 mm. long, 3 mm. broad. Petals 8, oblanceolate,
clawed, glabrous, 7–12 mm. long, (3–)4–7 mm. broad. Filaments 2–3 mm.
long; anthers pubescent, 3–4 mm. long. Ovary tomentose; style 4–5 mm.
long. Capsule subglobose, 1–1·5 cm. in diameter including the spines, the
latter inflated at the base and almost conical, pubescent, 3–6(–8) mm. long.
Seeds 2 or 3, pubescent, 6–7 mm. long, 5 mm. broad.

UGANDA. Bunyoro District: Budongo Forest, Nov. 1935, *Eggeling* 2288!; Masaka
  District: Minziro Forest, Oct. 1924, *Maitland* 1080!; Mengo District: Mukono,
  Oct. 1913, *Dummer* 304!
TANZANIA. Buha District: Kasakela Reserve, Melinda Stream, Nov. 1962, *Verdcourt*
  3374! & Feb. 1964, *Pirozynski* 370!
DISTR. U2, 4; T4; Cameroun, Sudan (Equatoria), Central African Republic (Haut
  Oubangui), Zaire
HAB. Lowland rain-forest and riverine forest; 600–1200 m.


4. **L. stipulata** (*Oliv.*) *Milne-Redh. & Sleumer* in F.R. 42: 260 (1937);
T.T.C.L.: 234 (1949); Sleumer in E.J. 94: 324 (1974). Type: Tanzania,
Tabora District, Jiwa la Mkoa, *Grant* (K, holo.!)

Multi-stemmed shrub, 1–3 m. tall; bark grey, smooth, slightly fissured.
Branchlets yellowish velutinous at tips, older parts glabrescent. Leaves
deciduous; blade broadly obovate, or oblong-, more rarely subovate-elliptic,
apex shortly or very shortly acuminate, base broadly attenuate to rounded,
or sometimes subcordate, recent ones softly yellowish tomentulose on both
faces, mature ones thin to chartaceous and ± glabrescent, still hairy mainly
on midrib and nerves on both faces with age, subsinuate-serrate to -dentate
with veins protruding beyond the edge, 5·5–8(–12) cm. long, 3–6(–8) cm.
broad; nerves 8–9(–10) pairs, rather straight from the midrib, prominent
beneath, reticulation lax, raised beneath; petiole tomentellous, 1–2 cm. long,
1·5–2 mm. in diameter; stipules persistent at least at the upper leaves, firm,
lanceolate-linear, pubescent, 5–7(–15) mm. long, 2(–3) mm. broad at base.
Inflorescences appearing with the new leaves, reduced to 1 or 2 flowers,
yellowish tomentellous; peduncle rather slender, 1·5–2 cm. long; pedicels
articulate at the middle, 1·5–5 cm. long. Sepals broadly elliptic, tomentose
outside, glabrous inside, 7–8 mm. long, 5–6 mm. broad. Petals 8–10(–12),
oblong-spathulate, shortly hairy outside, subglabrous inside, 12–14 mm.
long, 5–6 mm. broad. Filaments glabrous, 2–2·5 mm. long; anthers
pubescent, 3·5–4 mm. long. Ovary tomentose and muriculate; style hairy,
± 3 mm. long. Capsule ellipsoid-suborbicular, yellowish to orange, 1·2–
1·6 cm. long, 0·8–1 cm. in diameter, covered with numerous conical obtuse
emergences ± 1 mm. long, splitting into (2–)3(–4) valves. Seeds 1 or 2,
blackish, glossy, laxly hairy, 5–7 mm. long, 4–5 mm. broad.

TANZANIA. Mwanza District: Mbarika, Apr. 1953, *Tanner* 1379!; Kahama District:
  Mtwemi, May 1937, *B. D. Burtt* 5545!; Singida District: Itigi–Singida road, Mar.
  1965, *Richards* 19963!
DISTR. T1, 4, 5; not known elsewhere
HAB. *Brachystegia* woodland, deciduous bushland and thicket, often on rocky hills;
  1125–1500 m.

SYN. *Oncoba stipulata* Oliv. in Trans. Linn. Soc. 29: 31 (1873)
  *Buchnerodendron stipulatum* (Oliv.) Bullock in K.B. 1933: 468 (1933)
  *Lindackeria grewioides* Sleumer in F.R. 41: 121 (1936); T.T.C.L.: 234 (1949).
  Type: Tanzania, W. Dodoma, 456–460 km. on railway from Dar es Salaam,
  *Peter* 33144 (B, holo. †)

## 10. PHYLLOCLINIUM

Baillon in Bull. Soc. Linn. Paris 2: 870 (1890); Gilg in E. & P. Pf., ed. 2, 21:
431, fig. 196 (1925)

Shrubs. Leaves persistent, alternate, rather leathery, sinuate- to serrate-
dentate, petiolate, penninerved, stipulate. Inflorescence along the midrib
on the upper surface of the upper part of the leaves, either a few-flowered
cyme, or sometimes reduced to a solitary flower, subtended by a major leafy
bract and 2 lateral bracteoles; flowers bisexual or polygamous, pedicelled.
Sepals 3–4(–5), imbricate, rather irregular and dry. Petals 4–6, imbricate,
thin, much larger than the sepals. Stamens 25–40, free; filaments subulate;
anthers almost basifixed, dehiscent longitudinally, apiculate by the protruding connective. Ovary superior, 1-locular, with 2–4 multiovulate
placentas; style 1, shortly 2–4-fid. Capsule opening with 2 or 3 valves;
pericarp woody. Seeds albuminate.

A genus with 2 species in tropical Africa, one of them common in a restricted area
of western Tanzania.

**P. paradoxum** *Baillon* in Bull. Soc. Linn. Paris 2: 870 (1890); Gilg in E.J.
40: 502 (1908); Lecomte in Bull. Mus. Hist. Nat. Paris 24: 57, t. 1 (1918);
Exell & Mendonça in C.F.A. 1: 361 (1951); Pellegrin in Mém. Soc. Bot. Fr.
1952: 107 (1953); Cavaco, Ét. Fl. de la Lunda: 107 (1959); Bamps in
F.C.B., Flacourt.: 45 (1968); Letouzey, Hallé & Cusset in Adansonia, sér. 2,
9: 529 (1969). Type: Congo-Brazzaville, Loango area, *Thollon* 1343 (P,
holo.!)

Erect, not or little branched glabrous shrub, 0·5–2(–4) m. tall. Leaves
clustered near the tips of the branchlets; blade elongate-obovate or
-spathulate, apex rounded though tip abruptly acuminate for 1·5–2·5 cm.,
long-attenuate towards the base, subcoriaceous, ± sharply serrate-dentate,
15–36 cm. long, 4–10 cm. broad; lateral nerves 14–18 pairs, prominent on
both faces; petiole 1–10 cm. long; stipules lanceolate, 10–13 mm. long,
3–4 mm. broad, caducous. Inflorescence 1–3(–4)-flowered, on the midrib
of the upper ⅔ to ¾ of the lamina, glabrous; median bract ovate-lanceolate,
accrescent, (8–)10–40 mm. long, 6–20 mm. broad; lateral bracteoles 2–3 mm.
long. Pedicel 1–3(–10) mm. long. Sepals ovate, ciliolate, 5–9 mm. long,
3–4 mm. broad. Petals oblong, whitish, ciliate, 10–30(–40) mm. long,
5–10(–12) mm. broad. Filaments 4–8 mm. long; anthers 1·5 mm. long.
Ovary glabrous; style 5–8 mm. long, the 2–4 branches of the stigma ± 1 mm.
long. Capsule subglobose, ± 1 cm. in diameter; pericarp 1 mm. thick.
Seeds 2 or 3, subglobular-angular, ± 5 mm. in diameter, yellowish,
glabrous. Fig. 10.

TANZANIA. Kigoma District: Mukuyu, Nov. 1969, *Kielland* 86! & Mihumu near
  Lugufu R., Nov. 1971, *Kielland* 24! & Kasakati Basin, Mar. 1964, *Itani* 161!
DISTR. **T4**; Congo-Brazzaville, Zaire, Angola (Lunda)
HAB. In understorey of riverine forest; 900–1340 m.

SYN. *P. brevipetiolatum* Germain in B.J.B.B. 25: 262 (1955). Type: Zaire, Kasongo,
  Pene–Yumbe, *Germain* 7799 (BR, holo.!, L, iso.!)

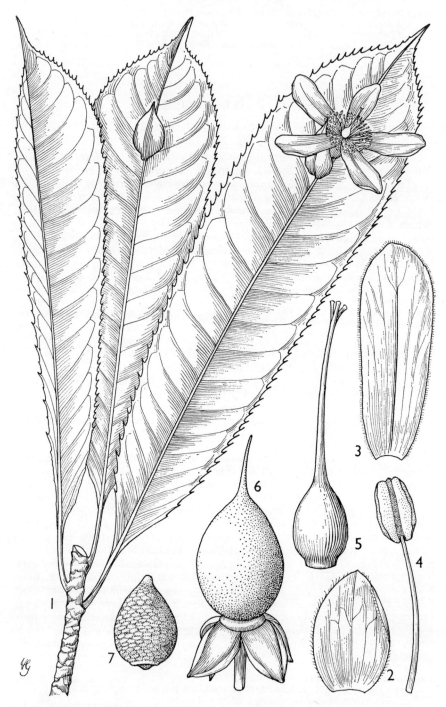

FIG. 10. *PHYLLOCLINIUM PARADOXUM*—**1,** flowering branchlet, × ⅔; **2,** sepal, × 4; **3,** petal, × 2; **4,** stamen, × 12; **5,** pistil, × 6; **6,** fruit, × 3; **7,** seed, × 4.  1, from *Kielland* 24 and *Itani* 30; 2, 4, 5, from *Kielland* 24; 3, from *Schouteden* 3; 6, from *Gossweiler* 13810; 7, from *Itani* 30.  Drawn by Victoria Goaman.

## 11. KIGGELARIA

L., Sp. Pl. 2 : 1037 (1753) & Gen. Pl., ed. 5 : 459 (1754)

Shrubs or trees with simple and stellate hairs. Leaves ± persistent, alternate, entire to serrate, petiolate, fugaciously stipulate. Flowers dioecious, the ♂ ones in axillary cyme-like short racemes, the ♀ usually solitary in the axils of the upper leaves. Sepals 5, almost free, ± valvate in bud. Petals 5, free, with a rather fleshy scale adnate to the base inside. Male flowers: stamens 8–12; filaments shorter than the anthers, the latter basifixed, the 2 thecae opening with terminal pores or slits. Rudiment of ovary 0. Female flowers: sepals and petals similar to those of the ♂ flowers. Ovary sessile, 1-locular, with 2–5 multi-ovulate placentas; styles 2–5, free or connate at base, short; stigmas obtuse. Fruit leathery to almost woody, dehiscing with 2–5 valves. Seeds 8–10, in a fleshy pulp, the testa surrounded by a viscid ± fleshy aril.

A monotypic genus from the mountains of Tanzania south to the Cape Peninsula.

**K. africana** *L.*, Sp. Pl. 2 : 1037 (1753); Harv. in Fl. Cap. 1 : 71 (1860); V.E. 3 (2) : 571 (1921); Gilg in E. & P. Pf., ed. 2, 21 : 413, fig. 179/F–H (1925); T.T.C.L. : 233 (1949); Wild in F.Z. 1 : 265, t. 43 (1960). Type: South Africa, Cape Province (LINN, holo. !)

Shrub or tree, up to 13 m. tall; bark pale to dark grey, smooth; branchlets ± densely stellately tomentellous. Leaf-blades variable in form, size, texture and indument-density, ovate-oblong, elliptic, or lanceolate, apex subacute or obtuse, base cuneate to rounded, papery to firmly chartaceous or subcoriaceous, yellowish stellate-tomentellous to -tomentose, sometimes glabrescent with age, rarely practically glabrous from the beginning except the nerves, entire, undulate or shallowly serrate, 3·5–9(–14) cm. long, 2–5(–6) cm. broad; lateral nerves 6–9 pairs, rather prominent beneath, veins ± parallel to each other, slightly raised; petiole 0·5–1·5(–2) cm. long. Flowers yellowish-greenish tomentellous, the ♂ in short 3–7-flowered axillary cyme-like racemes; peduncle ± 5 mm.; pedicels up to 7 mm.; basal bracts minute, caducous. Sepals narrowly ovate, slightly keeled, 4–5 mm. long and 2·5 mm. broad. Petals broadly obovate, obtuse or subacute, ± 5 mm. long, the basal inner scale oblong, fleshy, 1 mm. Filaments ± 1 mm.; anthers oblong, ± 2 mm. Female flowers solitary in the upper axils; peduncle ± 5 mm.; pedicels up to 1·5 cm. Sepals and petals similar, but slightly longer and narrower. Ovary tomentellous and minutely tuberculate; styles glabrous, ± 3 mm. long. Capsule globose, almost woody, ± 2 cm. in diameter, yellowish-greenish tomentellous or tomentose, softly warted. Seeds subglobose, covered with a bright orange-red aril; testa smooth, dark. Fig. 11.

TANZANIA. N. Kilimanjaro, May 1934, *Baldwin* 917!; Lushoto District: W. Shagayu [Shagai] Forest, May 1953, *Drummond & Hemsley* 2565!; Iringa District: Mt. Image, Mar. 1962, *Polhill & Paulo* 1627A!
DISTR. **T2, 3, 6, 7**; also in Malawi, Mozambique, Rhodesia and South Africa.
HAB. In and at margins of upland rain-forest and mist forest; 1370–2550 m.

SYN. *K. serrata* Warb. in P.O.A. C: 278 (1895); Gilg in E.J. 40: 468 (1908); V.E. 3 (2): 571 (1921); Gilg in E. & P. Pf., ed. 2, 21: 413 (1925); T.T.C.L.: 233 (1949). Types: Tanzania, Lushoto District, Kwambugu, *Holst* 3800 (B, syn. †) & Mazinde [Masinde], *Holst* 3883 (B, syn. †)
    *K. grandifolia* Warb. in P.O.A. C: 278 (1895); Gilg in E.J. 40: 468 (1908); V.E. 3 (2): 571 (1921); Gilg in E. & P. Pf., ed. 2, 21: 413 (1925); T.T.C.L.: 233 (1949). Type: Malawi, *Buchanan* 1469 (B, holo. †, K, iso. !)

FIG. 11. *KIGGELARIA AFRICANA*—**1,** branch with male flowers, × ⅔; **2,** male flower, with front sepals and petal removed, × 4; **3,** female flower, with front sepal and petal removed, × 4; **4,** fruits, × ⅔; **5,** seed, with part of aril removed, × 2; **6,** petal of male flower, × 4; **7,** petal of female flower, × 4. 1, 2, 6, from *Greenway* 6361; 3, 7, from *Codd* 6658; 4, 5, from *Worsdell*. Reproduced by permission of the Editors of Flora Zambesiaca.

*K. hylophila* Gilg in E.J. 40: 469 (1908); V.E. 3 (2): 571 (1921); Gilg in E. & P. Pf., ed. 2, 21: 413 (1925); T.T.C.L.: 233 (1949). Type: Tanzania, Uzungwa [Utschungwe] Mts., near Muhanga, *Goetze* 631 (B, holo. †)

*K. glabrata* Gilg in E. & P. Pf., ed. 2, 21: 413 (1925), *nom. nud.* Based on Tanzania, Rungwe District, Kondeland, *Stolz* 2137 (B, syn.†) & Njombe District, Madehani, *Stolz* 2478 (B, syn. †, K, iso.!)

*K. flavo-velutina* Sleumer in N.B.G.B. 12: 86 (1934); T.T.C.L.: 233 (1949). Type: Tanzania, Uluguru Mts., Lukwangule Plateau, *Schlieben* 3528 (B, holo. †, B, BM, K, P, iso.!)

## 12. SCOLOPIA

Schreb., Gen.: 335 (1789); Gilg in E. & P. Pf., ed. 2, 21: 418 (1925); Sleumer in Blumea 20: 26 (1972), *nom. conserv.*

Shrubs or trees, often with spines on the trunk and/or branches. Leaves alternate, persistent, subcoriaceous or coriaceous, entire or shallowly to rather deeply glandular-serrate-crenate, reddish or purplish when young, penninerved; stipules minute, caducous. Flowers small, bisexual, rarely also ♂ on the same specimen, in axillary, mostly simple, sometimes compound racemes, these rarely reduced to few-flowered fascicles or even to a solitary flower. Sepals 4–6, narrowly imbricate or subvalvate, ± connate at the base. Petals as many as sepals and similar with these. Receptacle flat, sometimes set with hairs around the base of the ovary and the base of the filaments. Extrastaminal disk, if present, composed of one row of free, short, thick orange glands. Stamens indefinite in number, pluriseriate, exceeding the petals at full anthesis; anthers dorsifixed, the connective often produced beyond the thecae into an apicular appendage. Ovary sessile, with 2–4(–5) few-ovuled placentas; style rather long, whether or not 3–4-partite distally; stigmas entire or slightly 2–3(–5)-lobed. Berry subglobose or ovoid, (1–)2–3(–6)-seeded, somewhat fleshy, with the withered sepals, petals and stamens at the base, crowned by the ± persistent style. Seeds with a hard testa, without an aril.

A genus of about 37 species, 6 confined to continental Africa, the rest in Madagascar, SE. Asia, Malesia and Australia, 4 being represented in East Africa.

Flowers sessile or subsessile in the axils of the leaves:
  Style abruptly demarcated from the developed
    ovary and/or young fruit; fruit oblong-ovoid .    1. *S. theifolia*
  Style gradually narrowing upwards from the
    ovary and/or young fruit; fruit ellipsoid-
    subglobular  .    .    .    .    .    2. *S. stolzii*
Flowers in few- to rather many-flowered axillary
  simple or compound racemes:
  Receptacle glabrous or moderately hairy; disk-
    glands absent; inflorescence rather lax-
    flowered; pedicels slender and accrescent .    3. *S. rhamniphylla*
  Receptacle ± densely white-hairy; disk-glands
    present; inflorescence rather compact; pedicels
    stoutish, hardly accrescent .    .    .    .    4. *S. zeyheri*

1. **S. theifolia** *Gilg* in E.J. 30: 359, fig. (1901) & 40: 484 (1908) & in V.E. 3 (2): 577, fig. 255 (1921) & in E. & P. Pf., ed. 2, 21: 420, fig. 188, as "*theiformis*" (1925); T.T.C.L.: 236 (1949); Sleumer in Blumea 20: 49 (1972). Type: Tanzania, Livingstone Mts., crest of Pikurugwe volcano, *Goetze* 1249 (B, holo. †, BM, BR, Z, iso.!)

Much branched shrub or tree, with rounded crown, unarmed, up to 15 m. tall, trunk up to 25 cm. in diameter; bark smooth, greyish white.

Leaf-blades elliptic to ovate-, oblong- or obovate-elliptic, apex broadly attenuate, obtuse, sometimes rounded, base cuneate, subcoriaceous, glabrous, edge regularly rather shallowly subserrate-crenate, (2·5–)3·5–6·5(–8) cm. long, (1·2–)1·7–2·7(–4) cm. broad; nerves in 4–7 pairs, lower 1 or 2 pairs steeply ascending, upper ones curved and more spreading, slightly though distinctly raised on both faces as is the rather dense reticulation; petiole initially puberulous, 2–5 mm. long. Flowers in the upper axils, generally 2–3(–4) in a fascicle, some sometimes solitary, sessile or subsessile, (3–)4–5-merous. Sepals ovate to suborbicular, ciliate, ± 2 mm. long. Petals similar to, though slightly smaller and thinner, than the sepals, sometimes minute, mostly caducous. Receptacle glabrous or practically so. Disk-glands numerous, subglobular. Stamens 30(–40); anthers hardly apiculate by the glabrous connective. Ovary glabrous; style 1–2 mm. long, stoutish, abruptly demarcated from the developed ovary and/or young fruit; stigma minute, subcapitate, obscurely 2–3-lobulate. Berry oblong-ovoid, ± 1 cm. long, 0·7 cm. across, red at full maturity, on a pedicel 1–2 mm. long. Seeds 1 or 2.

UGANDA. Karamoja District: Napak, June 1950, *Eggeling* 5979!
KENYA. Nakuru District: Rongai, Mar. 1932, *Cooper* in *F.D.* 2753! & Eastern Mau Forest Reserve, Sept. 1949, *Maas Geesteranus* 6208!; Kisumu-Londiani District: Tinderet Forest Reserve, June 1949, *Maas Geesteranus* 4918!
TANZANIA. Masai District: Longido Mt., Jan. 1936, *Greenway* 4391!; Lushoto District: Gologolo–Wite [Viti] road, Jan. 1952, *Parry* 106!; Mbeya Range, Feb. 1962, *Kerfoot* 3702!
DISTR. U1; K3–5; T2, 3, 7; Ethiopia, Sudan (Equatoria)
HAB. Upland dry evergreen forest and associated wooded grassland; 1600–2700 m.

SYN. *S. zavattarii* Chiov. in Miss. Biol. Borana, Racc. Bot.: 140 (1939). Type: Ethiopia, Boran District, *Cufodontis* 407 (FI, holo.!, W, iso.!)


2. **S. stolzii** *Gilg* in V.E. 3 (2): 577, in clavi (1921) & in E. & P. Pf., ed. 2, 21: 420 (1925); Sleumer in N.B.G.B. 12: 142 (1936); T.T.C.L.: 236 (1949); Wild in F.Z. 1: 278, t. 47/A (1960); F.F.N.R.: 267 (1962); Bamps in F.C.B., Flacourt.: 39 (1968); Sleumer in Blumea 20: 51 (1972). Type: Tanzania, Rungwe District, Kiwira R., *Stolz* 1742 (B, holo. †, BM, FHO, HBG, K, M, P, iso.!)

Much branched tree, 6–10(–15) m. tall, unarmed; bark pale brown, rather smooth, flaking. Leaf-blades elliptic, or subovate- to obovate-elliptic or -oblong, apex bluntly acuminate, or attenuate or rounded, base cuneate to rounded, coriaceous or subcoriaceous, glabrous, entire to regularly and shortly, sometimes remotely subserrate-crenate, (3–)5–7(–13) cm. long, (2–)3–5·5 cm. broad; nerves in 4–6 pairs, 2 basal pairs curved-ascending, upper ones shorter, reticulation finely prominent on both faces; petiole 2–7 mm. Flowers solitary in the upper axils, sessile or subsessile, 4–5(–7)-merous. Sepals ovate, ± 1·5 mm. long. Petals similar to, though more membranous and narrower than the sepals, ± caducous. Disk-glands numerous, subquadrate. Receptacle glabrous or hairy. Stamens 40–60; anthers very shortly appendiculate. Ovary glabrous or hairy; style ± 2 mm. long, shortly 3–5-partite distally. Fruit subglobose, fleshy, red, 2–2·5 cm. across, glabrous or sparsely hairy, on a stoutish pedicel 1–1·5 mm. long. Seeds 10–12.

### var. stolzii

Leaf-blades at apex bluntly acuminate, or rather obtuse, ± coriaceous, edge ± shallowly undulate, sometimes subserrate-crenate, (5–)7–10(–13) cm. long, (2–)3·5–5·5(–8) cm. broad. Ovary glabrous or slightly pubescent.

KENYA. Kiambu District: Thika, near base of Chania Falls, Dec. 1967, *Faden*!

Tanzania. Mbeya District: Itaka, Sept. 1933, *Greenway* 3656!; Rungwe District: Mulinda Forest, *Stolz* 1874!; Pemba I., Jambangome–Misufini, Dec. 1930, *Greenway* 2751!

Distr. **K**4; **T**7; **P**; Cameroun, Zaire, Zambia, Rhodesia, Malawi, Mozambique

Hab. In forest, forest fringe or riverine forest; up to 1500 m.

var. **riparia** (*Mildbr. & Sleumer*) *Sleumer* in Blumea 20: 52 (1972). Type: Tanzania, Ukinga Mts., Iletile R., *Stolz* 2200 (B, holo. †, L, M, W, Z, iso.!)

Leaf-blades at apex broadly attenuate to rounded, ± subcoriaceous, edge generally regularly and shallowly subserrate-crenate, 3–5(–7) cm. long, 2–3(–4) cm. broad. Ovary hairy.

Kenya. Masai District: Emali Hill, Dec. 1971, *Faden & Holland* 71/947!

Tanzania. Iringa District: Irundi, Nov. 1955, *Benedicto* 84! & Mufindi, Feb. 1932, *St. Clair-Thompson* 441!; Njombe District: Ruhudji R., Mpoponzi, July 1931, *Schlieben* 1092A!

Distr. **K**6; **T**7; Malawi

Hab. In dry evergreen, riverine forest or grassland with thickets; 1600–1950 m.

Syn. *S. riparia* Mildbr. & Sleumer in N.B.G.B. 11: 1077 (1934); T.T.C.L.: 235 (1949)

3. **S. rhamniphylla** *Gilg* in E.J. 40: 484 (1908) & in Z.A.E.: 568 (1913); R.E. Fries, Wiss. Ergebn. Schwed. Rhod.-Kongo-Exped. 1: 157 (1914); Gilg in V.E. 3 (2): 577 (1921) & in E. & P. Pf., ed. 2, 21: 420 (1925); F.P.N.A.: 636, t. 64 (1948); I.T.U., ed. 2: 150 (1952); K.T.S.: 229 (1961), as "*rhamnophylla*"; Bamps in F.C.B., Flacourt.: 42, t. 5 (1968); Sleumer in Blumea 20: 57 (1972). Type: Zaire, Lake Edward, *Scott Elliot* 8057 (B, holo.†, BM, iso.!)

Much branched shrub or tree, up to 12 m. tall, with rounded bushy crown; trunk sometimes armed with spines up to 15 cm. long; bark pale brown or grey, smooth or rough. Branches and branchlets with axillary straight spines up to 6 cm. long; sterile coppice shoots virgate, very spiny. Leaf-blades ovate-oblong or -lanceolate, or elliptic, or obovate, shortly acuminate, tip bluntish, base cuneate, subcoriaceous, initially puberulous on midrib and nerves, early glabrescent, 5–8(–12) cm. long, 2·5–4·5(–6) cm. broad, regularly subglandular-serrate-crenate; nerves (4–)5–8 pinnate pairs, slightly raised on both faces as is the dense reticulation; petiole 3–10 mm. long. Racemes from foliate and defoliate axils, solitary or rarely in pairs, rather few- and lax-flowered, sometimes reduced to fascicles, shortly pubescent or sometimes almost glabrous, 1–2(–4) cm. long; pedicels slender, 5–15 mm. long. Flowers 4–6-merous. Sepals ovate, puberulous outside, ± 2 mm. long. Petals linear, ± 2·5 mm. long, sometimes fugacious. Receptacle glabrous or moderately short-hairy. Disk-glands absent. Stamens 20–30; anthers hardly apiculate. Ovary glabrous; style 4–6 mm. long; stigma very shortly 3–4-lobed. Fruit ovoid-subglobular, finally red, 6–7 mm. across. Seeds few, angular, 3–4 mm. long. Fig. 12.

Uganda. Kigezi District: Mitano Gorge, Nov. 1950, *Purseglove* 3499!; Masaka District: Lake Nabugabo, Oct. 1953, *Drummond & Hemsley* 4729!; Mengo District: Kasa Forest, Dec. 1949, *Dawkins* 463!

Kenya. Kericho District: Cheptuiyet Forest Reserve, Mar. 1963, *Kerfoot* 4859!; Kwale District: Kwale, Apr. 1938, *Dale* in F.D. 3866! & Mrima Hill, Sept. 1957, *Verdcourt* 1896!

Tanzania. Bukoba District: Ruiga R. Forest, Apr. 1958, *Procter* 901!; Arusha District: Ngurdoto Crater National Park, Feb. 1966, *Greenway & Kanuri* 12384!; Iringa District: W. Mufindi, Jan. 1947, *Brenan & Greenway* 8262!

Distr. **U**2–4; **K**5, 7; **T**1–3, 6–8; Cameroun, Angola, Zaire, Rwanda

Hab. Rain-forest or dry evergreen forest and associated bushland, riverine forest, locally frequent; (100–)1000–2000 m.

Syn. *S. guerkeana* Gilg in E.J. 40: 483 (1908) & in V.E. 3 (2): 577 (1921); T.T.C.L.: 233 (1949). Type: Tanzania, Kilimanjaro, Kware [Quare] R. below Machame [Madschume], *Volkens* 2046 (B, holo. †, BR, iso.!)

FIG. 12. *SCOLOPIA RHAMNIPHYLLA*—**A**, habit, × ½; **B**, flower, × 5; **C**, longitudinal section of flower, × 5; **D**, transverse section of ovary, × 1½; **E**, fruit, × 3; **F**, seed, × 7. All from *Lebrun 5048*. Reproduced by permission of the Institut des Parcs Nationaux du Congo.

4. **S. zeyheri** (*Nees*) *Harv.* in Fl. Cap. 2: 584, in text (1862); Wild in F.Z. 1: 276 (1960); K.T.S.: 229 (1961); F.F.N.R.: 267 (1962); Bamps in F.C.B., Flacourt.: 40 (1968); Sleumer in Blumea 20: 60 (1972). Type: South Africa, Cape Province, van Staden Mts., *Ecklon & Zeyher* 1756 (B, holo. †, L, P, iso. !)

Much branched shrub or medium sized tree, occasionally up to 25 m. tall' often branching fairly low down; trunk up to 60 cm. in diameter, sometimes armed with branched spines; bark dark grey to brownish, rather thin. Younger plants with scandent habit, elongate branches and broader leaves. Branches and branchlets either unarmed, or often with strong axillary simple rather straight spines up to 20 cm. long, which sometimes bear leaflets and flowers. Leaf-blades variable in shape and size, ovate or obovate to oblanceolate, or rhomboid to elliptic, apex obtuse or rounded, rarely subacute, base cuneate, rarely obtuse or even cordate in coppice shoots, coriaceous or subcoriaceous, entire, repand or bluntly crenate, 2–8 cm. long, 1–3·5(–5) cm. broad; nerves 4–6 pairs at a rather narrow angle with the midrib, slightly raised on both faces as is the lax reticulation; petiole up to 1·5 cm. long. Racemes axillary, solitary, or rarely 2 or 3, few- to rather many-flowered, lax- to dense-flowered, shortly pubescent or partially glabrescent, 1–3(–6) cm. long; pedicels slender to stoutish, (1–)2–5(–10) mm. long. Flowers scented, (3–)4–5(–6)-merous. Sepals ovate, subacute, 1–1·5 mm. long, 1 mm. broad. Petals narrower than the sepals. Receptacle densely white-hairy. Disk-glands small, sometimes shortly pubescent. Stamens (20–)30–40; anthers shortly apiculate. Ovary glabrous; style stoutish, 1·5–2·5 mm. long; stigma obscurely 2–3-lobed. Fruit subglobular, fleshy, 7–8 mm. across, finally purple-red. Seeds 2 or 3, angular.

UGANDA. Ankole District: Rwampara, Mpororo Hill, Oct. 1932, *Eggeling* 653! & Ruizi R., Nov. 1950, *Jarrett* 314! & Lake Nakivali, Mar. 1931, *Kennedy* in *A.D.* 1984!
KENYA. Northern Frontier Province: Marsabit Mt., May 1958, *T. Adamson* 1!; Kiambu District: Muguga North, May 1961, *Verdcourt* 3131!; Masai District: Orengitok, May 1961, *Glover, Gwynne & Samuel* 1276!
TANZANIA. Bukoba District: Nshamba, Sept. 1935, *Gillman* 558!; Lushoto District: Shume Forest, May 1953, *Drummond & Hemsley* 2693!; Tanga District: Machui, Feb. 1955, *Faulkner* 1570!
DISTR. U2; K1, 3–7; T1–3, 6, 8; Cameroun, Angola, Rwanda, Mozambique, Zambia, Rhodesia, Botswana, Swaziland, in South Africa from Transvaal to the Cape Province
HAB. Dry evergreen forest, riverine forest and bushland, wooded grassland and open rocky or sandy sites, usually in drier places than *S. rhamniphylla*; 0–2400 m.

SYN. *Eriudaphus zeyheri* Nees in Eckl. & Zeyh., Enum. Pl. Afr. Austr. 2: 272 (1836)
*Scolopia cuneata* Warb. in P.O.A. C: 278 (1895); Gilg in E.J. 40: 481 (1908) & in V.E. 3 (2): 577 (1921). Type: Kenya, Mombasa, *Wakefield* (B, holo. †, K, iso. !)
*S. stuhlmannii* Warb. & Gilg in E.J. 40: 482 (1908); Gilg in V.E. 3 (2): 577 (1921); T.T.C.L.: 236 (1949), pro parte. Types: Tanzania, Lushoto District, Kwai–Gare, *Engler* 1202 (B, syn. †) & Uluguru Mts., Kibungu, *Stuhlmann* 8911 (B, syn. †)
*S. rigida* R.E. Fries in N.B.G.B. 9: 324 (1925); T.S.K.: 22 (1936); I.T.U., ed. 2: 150 (1952). Type: Mt. Kenya, W. side, *Fries* 1117 (UPS, holo.!, BR, K, S, iso.!)

## 13. GERRARDINA

Oliv. in Hook., Ic. Pl. 11, t. 1075 (1870)

Shrubs or small trees. Leaves persistent, alternate, petiolate, penninerved, stipulate. Flowers bisexual, in axillary pedunculate cymes; pedicel articulate in the lower third. Calyx campanulate, with a cupular receptacle and 5 unequal imbricate sepals, persistent. Petals 5, thin, small,

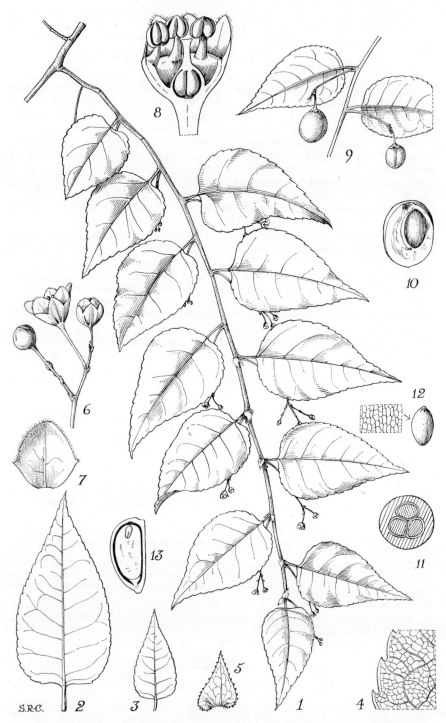

FIG. 13.  *GERRARDINA EYLESIANA*—**1,** flowering brachlet, × 1; **2, 3,** leaves, × 1; **4,** detail of leaf
margin and venation, × 4; **5,** stipule, × 4; **6,** inflorescence, × 4; **7,** petal, × 12; **8,** longitudinal
section of flower (petals removed), × 12; **9,** part of branchlet with fruits, × 1; **10,** longitudinal section of
1-seeded fruit, × 2; **11,** transverse section of 3-seeded fruit, × 2; **12,** seed, × 2, with detail of testa,
× 24; **13,** longitudinal section of seed, × 4.  Reproduced with permission of the Bentham-Moxon Trustees.

inserted on the margin of the disk, imbricate, alternating with the sepals, caducous. Disk perigynous, cupular, ± adnate to the receptacle. Stamens 5, opposite the petals, inserted on the margin of the disk; filaments subulate; anthers basifixed. Ovary at the deepened bottom of the receptacle, almost semi-inferior, 1-locular; placentas 2, each with 2 pendulous ovules; style short; stigma subcapitate or minutely bifid. Fruit dry or somewhat fleshy, indehiscent, 1–4-seeded. Seeds small; testa reticulate.

A genus of 2 species, one (*G. foliosa* Oliv.) limited to Natal and Swaziland, the other in eastern tropical Africa.

**G. eylesiana** *Milne-Redh.* in Hook., Ic. Pl. 34, t. 3390 (1939); Brenan in Mem. N.Y. Bot. Gard. 8: 219 (1953); Wild in F.Z. 1: 287, t. 50 (1960). Type: Rhodesia, Umtali, Stapleford Forest Reserve, *McGregor* 57/37 (K, holo.!, B, BM, EA, FHO, FI, W, iso.!)

Scandent shrub, up to 6 m. tall; branches slender, tips pubescent, glabrous below. Leaf-blades ovate to lanceolate, apex acutely acuminate, base broadly cuneate, truncate or cordate, thinly coriaceous, reddish when young, sparsely pubescent on the midrib beneath, otherwise glabrous, rather irregularly crenate-serrate, the teeth minutely red-glandular, 2–5·5 cm. long, 1·8–3 cm. broad; lateral nerves ± 7 pairs, prominent beneath, venation densely reticulate, markedly raised beneath; petiole 2–5 mm. long; stipules foliaceous, deltoid, apex acuminate, base truncate or slightly auriculate, glandular-serrate, up to 3 × 2 mm. Flowers in axillary pedunculate few-flowered cymes; peduncle slender, glabrous, or pubescent near the base, up to 2 cm. long; secondary branches short; pedicels 3–5 mm. long; bracts minute, ovate. Receptacle glabrous, ± 2 mm. Sepals deltoid-ovate, ciliolate, ± 2 mm. long, 1·5 mm. broad. Petals broadly ovate, ciliolate, white, 1·5–2 mm. long, 1·5 mm. broad. Stamens 1 mm. long; anthers apiculate. Disk cupular, ± adnate to the receptacle, obscurely pentagonal. Ovary ovoid, glabrous. Fruit pendulous, ellipsoid-ovoid, somewhat fleshy, ± 8 mm. long, 6 mm. across. Seeds 1–4, ellipsoid to plano-convex, ± 5 mm. long, 3 mm. broad; testa golden brown. Fig. 13.

TANZANIA. Lushoto District: W. Usambara Mts., Shagayu [Shagai] Forest, May 1953, *Drummond & Hemsley* 2532! & *Procter* 197; Kilosa District: Ukaguru Mts., S. of Mandege Forest Station, Mnyera ridge, 1 Jan. 1973, *Pócs & Lawrence* 6871/A.
DISTR. **T**3, 6; Rhodesia, Malawi, Mozambique
HAB. Margins or scandent in shrub layer of upland *Podocarpus-Ocotea* rain-forest; 2000–2100 m.

## 14. TRIMERIA

Harv., Gen. S. Afr. Pl., Suppl.: 417 (1838); Gilg in E. & P. Pf., ed. 2, 21: 429 (1925); Sleumer in E.J. 94: 302 (1974)

Shrubs or trees. Leaves persistent, alternate, (3–)5–9-nerved from the base, petiolate, glandular-serrate-crenate or -sinuate-dentate; stipules conspicuous, often foliaceous, finally caducous. Flowers dioecious, small, sessile, solitary or in glomerules along the rhachis of spike-like, simple or panicled axillary racemes. Sepals 3–5, slightly connate at the base, sub-imbricate. Petals 3–5, similar to the sepals, imbricate. Male flowers: disk-glands 3–5, opposite the sepals. Stamens 9–15, in bundles of 3 (rarely 4), alternating with the disk-glands; anthers minute. Rudiment of ovary minute. Female flowers: stamens 0. Ovary sessile, surrounded at the base by 3–5 flattish disk-glands, 1-locular, with 3 parietal placentas borne high up on the ovary-wall, each bearing 1–3 pendent ovules. Capsule 3-valved, subcoriaceous, 1–2(rarely–3)-seeded. Seeds arillate; testa crustaceous, finely pitted.

A genus of 2 species in tropical and subtropical, mainly central and ESE. Africa, 1 of which is represented in East Africa by a subspecies which is almost endemic there.

**T. grandifolia** (*Hochst.*) *Warb.* in E. & P. Pf., III. 6a : 37, fig. 13/H, J (1893) ; Sleumer in E.J. 94 : 305 (1974). Type: South Africa, Natal, Umlaas R., *Krauss* (B, holo. †, FI, M, W, Z, iso. !)

Shrub (rarely subscandent) or tree, 2–6(–20) m. tall ; bark grey to brownish. Branchlets pilose to densely soft-hairy to velutinous at tips, lower parts glabrescent. Leaf-blades ovate to elliptic or broadly reniform, apex shortly acuminate to obtuse or retuse, base rounded to cordate, firmly chartaceous to thinly coriaceous, densely soft-hairy to velutinous on both faces, sometimes glabrescent with age, particularly above, or almost glabrous, margin glandular-serrate-crenate, or more rarely -subsinuate-dentate, 2·5–8·5(–12) cm. long, 2·5–7·5(–12) cm. broad, 5–7(–9)-nerved from the base ; nerves raised mainly beneath, veins ± transverse, their network with the veinlets not much conspicuous ; petiole 0·8–1·5 cm. long ; stipules foliaceous, lanceolate to transversely elliptic, persistent for some time, up to 3–10 × 2–18 mm. Flowers small, sessile, densely pubescent, solitary or mostly in glomerules, these arranged along a slender rhachis, forming axillary spike-like racemes 2–5 (in the ♂ ones up to 9) cm. long. Sepals subovate-lanceolate, 1–1·5 mm. long. Petals subovate-elliptic, 1·5–2 mm. long. Male flowers : stamens 10–15 ; filaments thick-filiform, laxly patently hairy, 2–4 mm. long ; anthers ± 0·3 mm. across. Disk-glands 4 or 5, thick, subquadrate, glabrous. Female flowers : disk-glands 4, squamiform. Ovary elongate-ovoid, glabrous, 1 mm. ; styles 3, ± 1 mm. long. Capsule subtrigonous-ovoid, 2·5–3·5(–5) mm. long. Seeds few, ellipsoid, 2–3 mm. long.

subsp. **tropica** (*Burkill*) *Sleumer* in E.J. 94 : 308 (1974). Type : Tanzania, Tanga, Amboni Caves, *Holst* 2582 (K, holo. !, BM, iso. !)

Leaf-blades ovate or ovate-elliptic, rarely elliptic, apex generally shortly acuminate, rarely obtuse, often softly pubescent to velutinous ; stipules sometimes larger, up to 1 cm. long and 1·8 cm. wide. Inflorescences often densely pubescent. Fig. 14.

UGANDA. Ankole District : Ruizi R., Nov. 1950, *Jarrett* 140 ! ; Busoga District : Butembe-Bunya, Nov. 1937, *Webb* 44 ! ; Mubende Hill, Dec. 1937, *Sitwell* in *Eggeling* 3475 !
KENYA. Naivasha District : Karati R., May 1932, *Dale* in *F.D.* 617 ! ; Machakos District : Lukenya Hill, June 1953, *Bally* 8988 ! ; Kwale District : Shimba Hills, Feb. 1953, *Drummond & Hemsley* 1171 !
TANZANIA. Ngara District : Busubi, Jan. 1961, *Tanner* 5699 ! ; Arusha District : Mt. Meru, W. slope, Jan. 1936, *Greenway* 4408 ! ; Kondoa District : Bereku, Jan. 1928, *B. D. Burtt* 1139 !
DISTR. U1–4 ; K2–7 ; T1–3, 5, 6 ; E. Rhodesia, Rwanda, Burundi, Zaire (Lakes Albert, Edward and Kivu), Cameroun (Adamawa)
HAB. Dry evergreen or riverine forest, less often in rain-forest, often persisting in secondary growth, wooded grassland and bushland, sometimes on open rocky hillsides ; (150–)900–2450 m.

SYN. *T. tropica* Burkill in K.B. 1898 : 145 (1898) ; Gilg in E.J. 40 : 498 (1908) ; V.E. 1 (1) : 330, fig. 294 (1910) & 3 (2) : 582, fig. 258 (1921) ; Gilg in E. & P. Pf., ed. 2, 21 : 430, fig. 194 (1925) ; R.E. Fries in N.B.G.B. 9 : 324 (1925) ; T.S.K. : 24 (1936) ; T.T.C.L. : 550 (1949) ; K.T.S. : 500 (1961)
*T. macrophylla* Bak. f. in J.L.S. 37 : 154, t. 1 (1905). Type : Uganda, Ankole, near Mulema, *Bagshawe* 346 (BM, holo. !, K, iso. !)
*T. bakeri* Gilg in E.J. 40 : 499 (1908) ; V.E. 3 (2) : 583 (1921) ; R.E. Fries in N.B.G.B. 9 : 324 (1925) ; T.S.K. : 24 (1936) ; T.T.C.L. : 550 (1949) ; I.T.U., ed. 2 : 372 (1952) ; K.T.S. : 498 (1961). Type : Kenya, Nakuru/Masai District, Mau Plateau, *G. S. Baker* 24 (B, holo. †, EA, iso.)

NOTE. Transitions to subsp. *grandifolia* (Cape Province to Mozambique and E. Rhodesia) are found in E. Rhodesia, S. Tanzania (Ulanga District), and Mozambique (Manica).

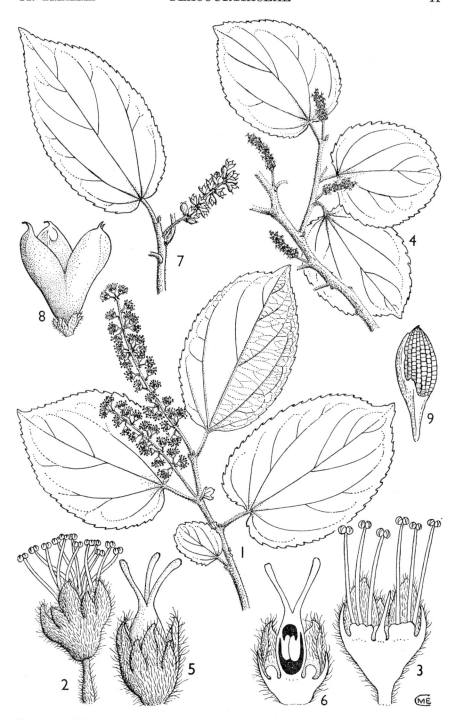

FIG. 14.   *TRIMERIA GRANDIFOLIA* subsp. *TROPICA*—**1,** male flowering branchlet, × ⅔; **2,** male flower, × 12; **3,** longitudinal section of male flower, × 14; **4,** female flowering branchlet, × ⅔; **5,** female flower, × 20; **6,** longitudinal section of female flower, × 20; **7,** fruiting branchlet, × ⅔; **8,** fruit, × 6; **9,** seed, ×9.  1–3, from *Kirrika* 215; 4–6, from *A. S. Thomas* 4192; 7–9, from *Greenway* 7864.   Drawn by Mrs. M. E. Church.

## 15. **HOMALIUM**

Jacq., Enum. Pl. Carib.: 5 (1760); Gilg in E. & P. Pf., ed. 2, 21: 425 (1925);
Sleumer in Bull. Jard. Bot. Nat. Belg. 43: 239 (1973)

Trees or shrubs. Leaves alternate, simple, entire or glandular-serrate-crenate, penninerved, petiolate; stipules absent, or minute to large, sometimes auriculate, caducous or ± persistent. Flowers bisexual, regular, (4–)5–8(–12)-merous, in axillary or subterminal racemes or panicles, solitary or fasciculate on the inflorescence-branches, sessile or pedicelled, subtended by small caducous or persistent bracts. Calyx-tube (or receptacle) adnate to the ovary. Sepals flat, usually narrow, persistent, often accrescent and wing-like. Petals alternating with the sepals, persistent, sometimes accrescent. Stamens epipetalous, solitary or in fascicles of 2–3; filaments slender; anthers small, extrorse, dorsifixed. Disk represented by a usually hairy gland opposite each sepal. Ovary conical and free in its upper half, sunk in the receptacle in its lower half, 1-locular, with 2–3(–8) placentas, each with 1–several ovules near the apex; styles 2–4(–7), free or connate to a distinct column below (then only stigmatic arms free); stigmas punctiform. Capsule half-inferior, coriaceous or woody, finally 2–3(–8)-valved from the apex, or remaining indehiscent (apparently mainly in sterile flowers). Seeds 1–few, small, often not fully formed.

A genus of about 180 species occurring throughout the tropics, with 5 species in East Africa.

Stamens in fascicles of 3 before each petal . . 1. *H. abdessammadii*
Stamens solitary before each petal:
  Flowers in simple, generally solitary racemes
    from the axils of the upper leaves:
    Leaves membranous, 5–8(–9) by (2–)2·5–3·5
      cm.; petioles very slender; rhachis very
      slender, few-flowered in its upper half . 2. *H. elegantulum*
    Leaves coriaceous or subcoriaceous, larger in
      general than above; petiole stoutish;
      rhachis ± robust and ± densely flowered
      for almost its entire length:
      Leaves with 4–5 pairs of lateral veins; flowers
        solitary, or rarely in pairs along the
        rhachis; lower part of the calyx-tube
        elongate-stipitate, i.e. ± twice as long
        as the upper obconical and ovuliferous
        part . . . . . . . 3. *H. gracilipes*
      Leaves with (6–)7–9(–10) pairs of lateral
        veins; flowers generally 2 or 3 per fascicle
        along the rhachis; lower part of the
        calyx-tube shortly stipitate, i.e. ± as long
        as the upper obconical and ovuliferous
        part . . . . . . . 4. *H. longistylum*
  Flowers in compound racemes forming panicles
    from the axils of the upper leaves . . 5. *H. africanum*

1. **H. abdessammadii** *Aschers. & Schweinf.* in Sitz. Ges. Nat. Freunde Berlin: 130 (1880); Hoffm. in Beitr. Kennt. Fl. Centr.-Ost-Afr.: 38 (1889); Gilg in E.J. 40: 494 (1908); V.E. 3 (2): 581 (1921); Gilg in E. & P. Pf., ed. 2, 21: 428 (1925); F.P.S. 1: 158 (1950); Fernandes & Diniz in Garcia de Orta 5 (2): 252 (1957); Wild in Bol. Soc. Brot., sér. 2, 32: 57

(1958), incl. ssp. *wildemanianum* (Gilg) Wild., & in F.Z. 1: 291, t. 51/A (1960); K.T.S.: 498 (1961); F.F.N.R.: 265 (1962); Sleumer in Bull. Jard. Bot. Nat. Belg. 43: 311 (1973). Type: Sudan, Equatoria, Ibba R. near Nganye, *Schweinfurth* 3954 (B, holo. †, K, P, iso. !)

Shrub or usually a tree, 5–10(–20) m., trunk straight, sometimes fluted, up to 40 cm. in diameter; bark smooth, greyish. Branchlets puberulous or glabrous, often purplish brown and with pale lenticels. Leaf-blade subcoriaceous, variable in shape and size, obovate or broad-elliptic, some-times lanceolate-ovate, ovate or almost rounded, entire or undulate, or ± deeply glandular-serrate-crenate, 5–10(–12) cm. long, (2·5–)3·5–5·5 cm. wide, subacuminate or obtuse, base broadly cuneate to subcordate, glabrous, or with (sometimes subsetular) hairs on midrib and secondary veins, and quite often in the nerve-axils beneath; lateral veins 6–7(–8) pairs, prominent beneath, reticulation rather coarse and but slightly raised beneath; petiole 0·7–1·5(–2) cm. long; stipules caducous. Spike-like racemes solitary or aggregated into subterminal or axillary sparsely branched and laxly flowered panicles up to 12 cm. long. Flowers subsessile, 5–7-merous, greenish-whitish, solitary or often 2 together. Calyx-tube broadly cup-shaped. Sepals puberulous on the back, elliptic or ovate, 3·5–5 mm. long. Petals broadly rhomboid-elliptic or ovate, puberulous dorsally, 3·5–4 mm. long. Stamens in fascicles of 3 opposite the petals; filaments glabrous or hairy, 3·5 mm. Disk-glands discoid, tomentose. Ovary conical, pilose; styles connate to a column below, stigmatic arms free and divergent. Capsule woody, similar in shape to the ovary, with persistent sepals and petals and usually containing 1 seed 2–3 mm. long and 1 mm. wide. Fig. 15.

UGANDA. Mbale District: Buginyanya, Dec. 1938, *A. S. Thomas* 2589!
KENYA. Kwale District: Gazi, Dec. 1936, *Dale* in *F.D.* 3586!; Kilifi District: Rabai–Ribe, *C. F. Elliott* 332!; Lamu District: Boni Forest, Oct. 1951, *Greenway & Rawlins* 9427!
TANZANIA. Lushoto District: Lwengera valley, Feb. 1960, *Semsei* 2985!; Tabora District: Ugalla R., July 1962, *Procter* 1952!; Uzaramo District: Msimbazi, Feb. 1940, *Vaughan* 2943!
DISTR. U3; K7; T3, 4, 6; westwards to Cameroun, Zaire and NE. Angola, northwards to the Sudan (Equatoria), southwards to Rhodesia, the Caprivi Strip, and Mozambique
HAB. Forest or forest edge, often in riverine forest; lowland to 1980 m.

SYN. *H. stuhlmannii* Warb. in E. & P. Pf., III. 6a: 36, fig. 14/E-F (1893). Type: Tanzania, Pangani, *Stuhlmann* I. 334 (B, holo. †, HBG, iso. !)
　　*H. boehmii* Gilg in E.J. 40: 494 (1908). Type: Tanzania, Mpanda District, Ugalla, Msima R., *Boehm* 89a in part (B, holo. †, Z, iso. !)
　　*H. warburgianum* Gilg in E.J. 40: 495 (1908). Type: Tanzania, Tabora District, Ugunda, Wala R., *Boehm* 89a in part (B, holo. †)
　　*H. macranthum* Gilg in E.J. 40: 495 (1908). Type: Mozambique, Niassa, Rovuma R. opposite Mt. Lissenga, *Busse* 1049 (B, holo. †, EA, iso.)
　　*H. wildemanianum* Gilg in E.J. 40: 497 (1908). Type: Zaire, Katanga, Lukafu, *Verdick* 123 (B, lecto. †, BR, isolecto. !) & 130 (B, syn. †, BR, isosyn. !)
　　*H. setulosum* Gilg in E.J. 40: 497 (1908). Type: Zaire, Bas-Zaire, Luozi cataracts, *Luja* 128 (B, holo., cit. " 428 " †, BR, iso. !)
　　*H. eburneum* Engl., V.E. 3(2): 581 (1921), in obs. Type: Cameroun, Molundu, *Mildbraed* 3911 (B, holo. †, HBG, iso. !)
　　*H. rhodesicum* Dunkley in K.B. 1934: 184, fig. (1934). Type: Zambia, Southern Province, Kafue R., *Martin* 66 (K, holo. !, FHO, iso. !)
　　[*H. rufescens* sensu Miller in Journ. S. Afr. Bot. 18: 59 (1952), *non* (E.Mey.) Benth.]

2. **H. elegantulum** *Sleumer* in N.B.G.B. 12: 715 (1935); T.T.C.L.: 549 (1949); Sleumer in Bull. Jard. Bot. Nat. Belg. 43: 269 (1973). Type: Tanzania, Lindi District, Noto Plateau, *Schlieben* 6111 (B, holo. †, B, HBG, iso. !)

FIG. 15. *HOMALIUM ABDESSAMMADII*—**1**, fertile branch, × ⅔; **2**, flower, × 4; **3**, sepal, × 8; **4**, petal, × 8; **5**, stamen, × 18; **6**, pistil, × 6; **7**, fruit, × 4; **8**, seed, × 4. 1–6, from *Bullock* 3043; 7, 8, from *Procter* 2090. Drawn by Victoria Goaman.

Shrub, 1–2 m. high; branchlets slender, tips puberulous. Leaf-blade oblong to obovate-oblong, obtusely attenuate, base broadly cuneate, membranous, glabrous except midrib, obtusely subserrate-crenate, 5–8(–9) cm. long, (2–)2·5–3·5 cm. wide, with 4–6 curved pairs of lateral veins, a little prominent on both faces as is the reticulation; petiole very slender, 4–5 mm. long. Racemes spike-like, greyish puberulous all over; rhachis very slender, (5–)7–11 cm. long, bearing distant solitary flowers or fascicles of 2–3 flowers in the upper part only; pedicels slender, 1–2 mm. long. Calyx-tube turbinate, ± 2·5 mm. long, attenuate at base to a short slender stipe. Sepals 6–7, linear-oblong, ± 2 mm. long. Petals 6–7, broadly oblong, pale yellow, sericeous on both sides, ± 3 mm. long. Stamens single before each petal; filaments glabrous or nearly so. Disk-glands roundish and flattish, glabrous or nearly so. Ovary shortly conical, sericeous below, glabrous above and on the columnar style (2 mm.). Fruit not known.

TANZANIA. Lindi District: Noto Plateau, Mar. 1935, *Schlieben* 6111!
DISTR. T8; apparently rare, not yet recollected
HAB. Scrub forest; 450 m.

3. **H. gracilipes** *Sleumer* in Bull. Jard. Bot. Nat. Belg. 43: 269 (1973). Type: Tanzania, Songea District, Matengo Hills, Liwiri [Luwira]-Kiteza Forest Reserve, *Semsei* 2538 (K, holo.!, BR, EA, FHO, iso.!)

Tree up to 27 m. high. Branchlets glabrous, soon covered with pale elliptic lenticels. Leaf-blade persistent, elliptic or broad-oblong, shortly obtusely acuminate, base cuneate to the petiole, subcoriaceous, glabrous, shining, subentire to very shallowly glandular-crenate, 5–8 cm. long, (2–)2·5–4 cm. wide; lateral veins in 4(–5) curved pairs, raised on both faces as is the dense reticulation; petiole 3–5 mm. long. Racemes solitary from the axils of the upper leaves, rather densely flowered in their upper ⅔ part, 6–9 cm. long; rhachis slender, subglabrous. Flowers solitary or 2 in a fascicle, 6(–7)-merous, white, shortly pubescent. Pedicels very slender, 1·5–2·5(–4) mm. long. Calyx-tube funnel-shaped in the upper ovuliferous part for 1·5–2·5 mm., rather abruptly narrowed downwards to a pedicel-like slender stipe for 4–5 mm. Sepals narrowly deltoid, subacute, 2 mm. long, hardly accrescent. Petals oblong-spathulate, puberulous on both faces, ciliolate, 6–8 mm. long and ± 2 mm. wide after flowering (perhaps a little longer in mature fruit). Filaments subglabrous or laxly patently hairy. Glands laxly hairy. Upper conical part of the ovary densely hairy; styles 3, connate for 1–1·5 mm. and hairy below, free for 1–1·5 mm. distally and glabrous but for the tips.

TANZANIA. Songea District: Liwiri [Luwira]-Kiteza Forest Feserve, Oct. 1956, *Semsei* 2538!
DISTR. T8; not yet recollected
HAB. In forest; ± 1900 m.

4. **H. longistylum** *Mast.* in F.T.A. 2: 497 (1871); Sleumer in Bull. Jard. Bot. Nat. Belg. 43: 270 (1973). Type: Gabon, Corisco I., *Mann* 1855 (K, holo.!)

Small to high tree; bark hard, greyish or whitish, rough. Branchlets glabrescent. Leaf-blade ovate- or oblong-elliptic, shortly, subabruptly and obtusely acuminate, base broadly cuneate to rounded, coriaceous or sub-coriaceous, glabrous except sometimes for hairs in the angles of the lateral veins with the midrib beneath (domatia), entire, undulate or coarsely crenate, 7–14 cm. long, 4–6·5 cm. wide; lateral veins 7–9 pairs, raised on both faces as is the rather dense reticulation; petiole 6–10 mm. long.

Racemes solitary from the axils of the upper leaves, 10–25 cm. long, rather densely flowered, forming together a terminal pseudopanicle when the leaves are fallen in later stages; rhachis rather slender, puberulous as are the 2–3-fascicled whitish-greenish 5(–6)-merous flowers. Calyx-tube at anthesis 1·5–2 mm. long, narrowed downwards to a slender stipe 1–2 mm. long. Sepals subtriangular-lanceolate, 1·5–2 mm. long, a little accrescent. Petals oblong-spathulate, puberulous on both faces and ciliolate, (3–)6–8 mm. long in full anthesis, accrescent to 10–15 × 3–4 mm. in fruiting stage. Filaments glabrous. Glands hairy. Styles 5(–6), connate into a hairy column for 1·5–2 mm., free and glabrous for ± 1 mm. distally.

KENYA. Kilifi District: Rabai, Nov. 1933, *Joanna* in *C.M.* 5949 !
TANZANIA. Lushoto District: Derema, Jan. 1899, *Scheffler* 214 !
DISTR. **K7**; **T3**; westwards to western and central Africa, southwards to Zambia and Mozambique
HAB. Forest or forest edge, or riverine forest; 600–800 m.

SYN. *H. calodendron* Gilg [in N.B.G.B. 3 : 84 (1900), nom. nud.] in E.J. 40 : 491 (1908).
Type: Tanzania, Derema, *Scheffler* 214 (B, holo. †, BM, BR, EA, K, P, Z, iso. !)
*H. mossambicense* Paiva in Bol. Soc. Brot., sér. 2, 40 : 266, t. 2, 3 (1966). Type: Mozambique, Niassa, Cabo Delgado, Mocimba da Praia–Palma road, *Gomes e Sousa* 4603 (COI, holo., EA, K, iso. !)

5. **H. africanum** (*Hook. f.*) *Benth.* in J.L.S. 4 : 35 (1859); Mast. in F.T.A. 2 : 497 (1871), pro parte; Wild in F.Z. 1 : 290 (1960); F.F.N.R.: 266 (1962); Sleumer in Bull. Jard. Bot. Nat. Belg. 43 : 281 (1973). Type: Sierra Leone, *G. Don* (BM, holo. !)

Shrub or generally a tree up to 25 m. high; trunk up to 40 cm. in diameter; bark greyish-whitish or brownish, rather smooth, hard. Branchlets pubescent or glabrous. Leaf-blade variable in size, shape, texture and degree of pubescence, generally oblong or elliptic-oblong, (8–)10–18(–26) cm. long, 4–7(–12) cm. wide, apex ± abruptly acuminate, rarely blunt, base cuneate to rounded, or rarely cordate, ± coriaceous, glabrous or shortly hairy, often hairy in the nerve-axils beneath, usually ± coarsely serrate-crenate, rarely ± entire, with (7–)8–10(–12) pairs of lateral veins, these curved and ± prominent beneath, reticulation obvious but often but slightly prominent; petiole ± robust, 6–12(–15) mm. long; stipules persistent (especially in sterile shoots) or caducous, linear or foliaceous (falcate to auriculate), up to 2·5 cm. long. Panicles ± interruptedly dense-flowered, up to 30 cm. long and 25 cm. wide, generally grey puberulous all over. Flowers greenish white, 2–5-fascicled along the rhachis, 5–6(–7)-merous. Pedicels 0·5–2(–2·5) mm. Calyx-tube obconical, attenuate to a short rather slender stipe. Sepals lanceolate, 1·5–2 mm. long. Petals oblanceolate-spathulate, sometimes a little fleshy or swollen, densely shortly hairy on both faces, in fruit accrescent to 4 mm. Filaments glabrous or laxly hairy below, 2(–3) mm. long. Glands cushion-like, almost reniform, grey puberulous. Ovary conical, densely hairy; styles (3–)4–5, connate to a column for 1–2 mm. below. Capsule not longer than the ripe ovary. Seeds usually single.

TANZANIA. Mpanda District: Kungwe-Mahali Peninsula, Mugombazi R., Aug. 1959, *Harley* 9466 !; Morogoro District: Ruvu R., *Stuhlmann* 8938; Njombe District: Mugwe R. valley, Nov. 1931, *Schlieben* 1447 !; Pemba I., Chake Chake, Oct. 1929, *Vaughan* 689
DISTR. **T4, 6, 7**; **P**; from Guinée to Angola, Zaire, Malawi, Zambia, Mozambique
HAB. Riverine forest, also in transitional drier forest; (0–)600–1300 m.

SYN. *H. sarcopetalum* Pierre in Bull. Soc. Linn. Paris 1 : 119 (1899); Wild in F.Z. 1 : 290 (1960). Types: Gabon, Libreville, *Klaine* 337 (P, lecto. !, BM, BR, isolecto. !) & 672 (P, syn. !)

[*H. stipulaceum* sensu Engl. in E.J. 28: 438 (1900), *non* Mast.]

*H. molle* Stapf in J.L.S. 37: 100 (1905). Type: Liberia, Sinoe Basin, *Whyte* (K, holo.!)

*H. riparium* Gilg in E.J. 40: 494 (1908); V.E. 3 (2): 580 (1921); Gilg in E. & P. Pf., ed. 2, 21: 427 (1925); T.T.C.L. 2: 549 (1949); Fernandes & Diniz in Garcia de Orta 5 (2): 249, 251, t. 5 (1957). Type: Tanzania, Morogoro District, Ruvu R., *Stuhlmann* 8938 (B, lecto.†) & Uluguru Mts., Kidai, *Stuhlmann* 9004 (B, syn. †)

*H. molle* Stapf var. *rhodesicum* R.E.Fries, Wiss. Ergebn. Schwed. Rhod.-Kongo-Exped. 1: 156 (1914). Type: Zambia, Mporokoso District, Kunkuta, *Fries* 1180 (UPS, holo.!)

NOTE. *H. africanum* is conceived here in a broad sense. Widely distributed in tropical Africa, it comprises numerous local forms or ecotypes which, as far as described as proper species, are regarded here as synonyms. Such species mentioned for eastern Africa are listed above.

## 16. CASEARIA

Jacq., Enum. Pl. Carib.: 4 (1760); Gilg in E. & P. Pf., ed. 2, 21: 451 (1925)

Shrubs or mostly trees, sometimes up to 40 m. tall. Leaves alternate, distichous, coriaceous or rarely chartaceous, entire, wavy, or subserrate-crenulate especially when young, pellucid-punctate and/or -lineate at least in the young state, penninerved; lateral veins arcuate, or rarely nearly straight from the midrib; stipules small, caducous. Flowers bisexual, small, solitary or usually few to numerous, clustered in axillary or slightly supra-axillary fascicles or glomerules, each from a cushion formed by few to numerous small scale-like bracts, ⚥. Pedicels articulated at the base. Receptacle cupular or funnel-shaped, ± perigynous, ± deeply 5-lobed; calyx-lobes slightly imbricate, persistent. Petals 0. Stamens 8–10(–12), equal or alternate ones with longer filaments, alternating with as many flattened or clavate, usually densely hairy appendages (or staminodes) and connate with them below into a ± perigynous tube; filaments filiform; anthers small, often apiculate by the slightly protruding connective. Ovary ovoid, attenuate to a short style; stigma capitate to almost disk-like. Capsule coriaceous, hard or succulent in its outer part, yellow to orange-red, ovoid-globose to ellipsoid, slightly 3–4(–6)-angled when fresh, often 4–6-ribbed when dry, splitting from above into (2–)3–4 valves. Seeds few to numerous, partly enveloped in a soft membranous, usually fimbriate aril which becomes reddish when exposed to the air.

A pantropical genus of about 160 species, of which only 12 are represented in all tropical Africa, Madagascar, and the Mascarenes, with 4 species in East Africa.

The species accepted here are close to each other, and mainly distinguished by leaf-characters; they show, as far as can be said with the still rather sparse materials at hand, a pattern of distribution comparable to genera in many other families within Africa.

Leaves moderately prominent-reticulate beneath only, upper surface smooth, edge entire, flat; petiole 10–13(–15) mm.; flowers glabrous; fruit (1·5–)2–2·5(–3) cm. long . . . . . . 1. *C. runssorica*

Leaves prominent-reticulate on both faces (though often more distinctly so beneath):

Leaves stiffly coriaceous, entire, edge generally revolute in dry specimens; petiole 3–5(–6) mm.; flowers glabrous; fruit 2·5–3(–3·5) cm. long . 2. *C. engleri*

Leaves ± flexibly subcoriaceous to coriaceous, edge flat, entire, or undulate, or subserrate-crenulate especially when young; petiole (6–)10–15(–18) mm. long; flowers pubescent externally; fruit 1–1·8 cm. long:

Leaves subcoriaceous to coriaceous, entire or
   undulate in the mature state; anthers glabrous   3. *C. gladiiformis*
Leaves firmly chartaceous to subcoriaceous, often
   still subserrate-crenulate in the mature state;
   anthers very shortly hairy . . . . 4. *C. battiscombei*

1. **C. runssorica** *Gilg* in Z.A.E.: 570 (1913); V.E. 3 (2): 589 (1921);
Gilg in E. & P. Pf., ed. 2, 21: 454 (1925); Sleumer in B.J.B.B. 41: 421
(1971). Type: Zaire, W. Ruwenzori, Butagu R., *Mildbraed* 2676 (B, holo. †)

Small or middle-sized tree, rarely up to 40 m. high, bole cylindric,
20–40 (rarely up to 60) cm. in diameter; bark greyish-brownish, slightly
rugose; branches ± horizontal; branchlets prismatic-angular and glabrous
at tips. Leaf-blade coriaceous, ovate-oblong or oblong, entire, (5–)7–
14(–18) cm. long, (3–)4–6(–8) cm. wide, subacuminate, broadly cuneate or
sometimes almost rounded at the usually only slightly unequal-sided base,
smooth and shining above; lateral veins 6–7(–9) pairs, arcuate, prominent
beneath only; tertiary veins laxly prominent-reticulate beneath; petiole
1–1·3(–1·5) cm. long. Inflorescences axillary, fascicled or glomerate, from
a subglobose many-bracteolate cushion, glabrous; pedicels 3(–5) mm. long.
Flowers greenish. Calyx-lobes ± 3 mm. long. Fruit subglobular to broadly
ellipsoidal, subtrigonous, yellow-orange, few-seeded, (1·5–)2–2·5(–3) ×1·5(–2)
cm.

UGANDA. Bunyoro District: Budongo Forest, May 1935 & Nov. 1937, *Eggeling* 1728!;
   Kigezi District: Kanungu, Oct. 1940, *Eggeling* 4194!; Masaka District: Sese Is.,
   Sozi, June 1925, *Maitland* 788!
TANZANIA. Bukoba District: Minziro Forest Reserve, Sept. 1950, *Watkins* 534 in *F.D.*
   3273! & Kantale, *Gillman* 383! & without locality, *Gillman* 622!
DISTR. U2, 4; T1; S. Sudan, NE. and east-central Zaire, Rwanda, Burundi
HAB. Rain-forest and semi-swamp forest; (750–)850–2100(–2450) m.

SYN. [*C. engleri* sensu Andrews, F.P.S. 1: 158 (1950); I.T.U., ed. 2: 372 (1952), *non*
   Gilg]

2. **C. engleri** *Gilg* in E.J. 40: 511 (1908); V.E. 3 (2): 590 (1921); Gilg in
E. & P. Pf., ed. 2, 21: 454 (1925); T.T.C.L.: 548 (1949); Sleumer in
B.J.B.B. 41: 410 (1971). Type: Tanzania, Lushoto District, Mbalu,
*Engler* 1446 (B, holo. †)

Tree up to 20 m. high, bole erect; bark smooth, pale grey; branchlets
striate, glabrous. Leaf-blade stiffly coriaceous, obovate or oblong, entire,
edge usually ± revolute in dry specimens, 6–11(–14) cm. long, 3–5·5(–7) cm.
wide, generally distinctly unequal-sided, shortly attenuate or subacuminate,
base rounded or rarely subcordate, or mostly broadly cuneate on one side
and rounded on the other; lateral veins 8–10 pairs, arcuate, more distinctly
raised beneath; tertiary veins forming a dense network slightly though
manifestly prominent on both faces; petiole 3–5(–6) mm. long. Inflorescences axillary, fascicled or glomerate, from rather small many-bracteolate
cushions, glabrous; pedicels 1–2 mm. long. Flowers greenish. Calyx-lobes
3 mm. long. Fruit ovoid-ellipsoidal, subacutely attenuate, longitudinally
6-angular, pale yellow to orange, with a characteristic smell, few-seeded,
2·5–3(–3·5) × ± 1·5 cm.

TANZANIA. Lushoto District: W. Usambara Mts., 2 km. NW. of Kwai, June 1953,
   *Drummond & Hemsley* 2904! & Shume Forest, June 1932, *Wigg* 136! & near
   Shagayu Sawmill, Sept. 1959, *Semkiwa* 59!
DISTR. T3; endemic
HAB. Upland rain-forest and dry evergreen forest; 1370–2135 m.

3. **C. gladiiformis** *Mast.* in F.T.A. 2 : 493 (1871) ; Warb. in E. & P. Pf., III.
6a : 51 (1893); Engl., P.O.A. C: 279 (1895); Gilg in E.J. 40: 510 (1908);
V.E. 3 (2): 589 (1921); Gilg in E. & P. Pf., ed. 2, 21: 454 (1925);
Fernandes & Diniz in Garcia de Orta 5: 252 (1957); Wild in F.Z. 1: 293,
t. 52/B (1960); Sleumer in B.J.B.B. 41: 423 (1971). Type: Mozambique,
Shupanga [Lacerdonia], *Kirk* (K, holo. !)

Large shrub or tree up to 15 m. tall. Bark greyish. Branchlets ± densely
short-hairy, glabrescent. Leaf-blade (5–)10–18 cm. long, (2·8–)3·5–7 cm.
wide, lanceolate or oblong, or subovate-oblong to -elliptic, shortly
acuminate or obtuse, base ± oblique, cuneate, or the smaller (sometimes
more highly inserted) side broadly cuneate, and the wider one almost
rounded, subcoriaceous to coriaceous, firm, glabrous, or fugaciously
puberulous on the midrib beneath, with circular and linear pellucid dots,
somewhat shining above, entire or undulate, or when young irregularly
subserrate-crenulate, with (7–)8–10 pairs of curved lateral veins, these
prominent mainly beneath, reticulation rather dense and distinctly raised
on both faces; petiole (7–)10–15(–18) mm. long. Flowers greenish white,
in dense axillary fascicles or glomerules on a cushion of minute puberulous
bracts; pedicels rather slender, puberulous, (1–)2–4 mm. long. Calyx-lobes
shortly hairy dorsally or glabrescent, (2–)3 mm. long. Filaments glabrous,
or very slightly short-hairy. Anthers glabrous. Fruit ovoid-subglobular
or ellipsoid, slightly 5–6-angular, splitting from above into 3–4 longitudinal
valves, (1–)1·2–1·6(–1·8) cm. long, 0·9–1·2 cm. in diameter. Fig. 16/B.

KENYA. Mt. Kenya, *Gardner* in *F.D.* 1881 !; N. Kavirondo District: Bukura, Dec.
1943, *M. D. Graham* 45 ! & Kakamega Forest, Sept. 1949, *Maas Geesteranus* 6265 !
TANZANIA. Morogoro District: Mtibwa Forest Reserve, Dec. 1953, *Semsei* 1528 !;
Uzaramo District: Pugu Hills, Sept. 1964, *Procter* 2754 !; Rufiji District: Mafia I.,
Oct. 1937, *Greenway* 5363 !; Zanzibar I., Ufufuma, Dec. 1932, *Vaughan* 2035 !;
Pemba I., Ngezi forest, Jan. 1933, *Vaughan* 2067 !
DISTR. **K**3–5; **T**3, 6; **Z**; **P**; Malawi, Mozambique, South Africa (Natal)
HAB. In rain-forest, dry evergreen, riverine and secondary forest, also in coastal
woodland and bushland; 0–600(in Kenya –1700) m.

SYN. *C. junodii* Schinz in Mém Herb. Boiss. 10: 52 (1900). Type: Mozambique,
      Delagoa Bay, *Junod* 351 (Z, holo. !, BR, G, K, P, iso. !)
    *C. macrodendron* Gilg in E.J. 40: 510 (1908). Types: Tanzania, Uzaramo,
      *Stuhlmann* 8599 (B, lecto. †) & Uluguru Mts., Tawa [Tana], *Stuhlmann* 8924
      (B, syn. †)
    *C. holtzii* Gilg in E.J. 40: 510 (1908). Types: Tanzania, Uzaramo District:
      Pugu Hills, *Holtz* 649 & 659 (B, syn. †)
    [*C. engleri* sensu Battiscombe, T.S.K.: 24 (1936); K.T.S.: 498 (1961), pro parte,
      *non* Gilg]

NOTE. The specimens from Kenya with their more coriaceous leaves are not typical.
The leaves of *C. gladiiformis* are smaller in drier localities, mainly towards the south
of the area of distribution.

4. **C. battiscombei** *R.E. Fries* in N.B.G.B. 9: 326 (1925); T.S.K.: 23
(1936); I.T.U., ed. 2: 372 (1952); Wild in F.Z. 1: 294, t. 52/A (1960);
K.T.S.: 497 (1961); Sleumer in B.J.B.B. 41: 419 (1971). Type: Mt. Kenya,
Embu District, Kiye R. [Kii], *Fries* 2012 (UPS, holo. !, K, iso. !)

Small or medium-sized tree with horizontally straggling branches, or
generally a tall forest tree up to 40 m. high and with straight, sometimes
fluted bole 30–45(–90) cm. in diameter and a vertical crown; bark rough,
grey to brownish, or almost dark on old trees, peeling off in ± rectangular
thin patches. Branchlets usually puberulous, glabrescent, silvery in saplings.
Leaf-blade oblong or narrowly oblong, (8–)12–22 cm. long, 3–5(–7) cm.
wide, generally rather unequal-sided, shortly obtusely acuminate or obtuse,
base cuneate to rounded on one, and rounded on the other side, or

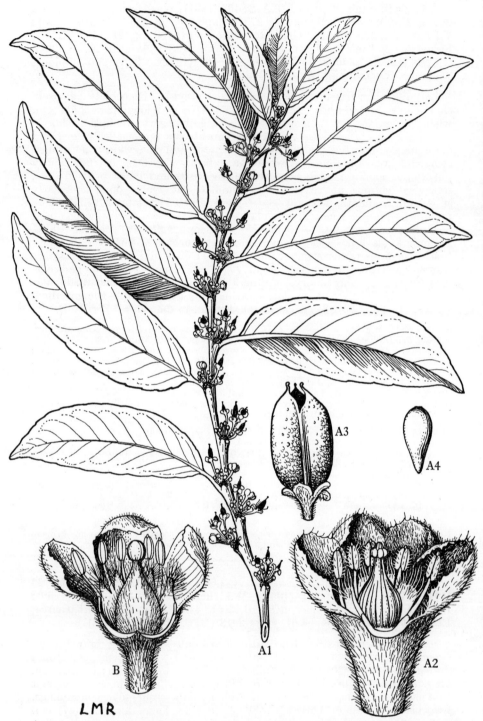

FIG. 16. *CASEARIA BATTISCOMBEI*—**A**1, flowering branch, × ⅔; **A**2, flower, with calyx-lobe, one stamen and one appendage removed, × 14; **A**3, fruit, × 2; **A**4, seed, × 2. *C. GLADIIFORMIS*—**B**, flower, with two calyx-lobes, five stamens and five appendages removed, × 6. A1, 3, 4, from *Battiscombe* 19; A2, from *Wild* 2239; B, from *Pedro* 4194. Reproduced by permission of the Editors of Flora Zambesiaca.

subcordate, glabrous, or fugaciously hairy beneath, firmly chartaceous to subcoriaceous, subentire or (especially on saplings) subserrate-crenulate or undulate, and with (12–)14–20 pairs of rather irregular curved secondary veins, these ± prominent beneath, reticulation but slightly raised beneath; petiole 6–10(–15) mm., glabrous or puberulous. Flowers numerous, hairy, greenish-yellowish, in axillary fascicles borne on a cushion of minute pubescent bracts; pedicels (2–)3–4 mm. Calyx-lobes ± 2 mm. long. Filaments and anthers very shortly hairy. Fruit subovoid-ellipsoid, slightly angular, 2–4-valved, yellow-orange, 1–1·3 cm. long, ± 0·8 cm. in diameter. Seeds few. Fig. 16/A.

UGANDA. Ankole District: Kalinzu Forest Reserve, May 1956, *Katentera* 1!; Elgon, Feb. 1940, *St. Clair-Thompson* in *Eggeling* 3954! & 3955!
KENYA. Northern Frontier Province: Mt. Kulal, July 1958, *Verdcourt* 2258!; Nandi District: July 1913, *Battiscombe* 664!; Nyeri Hill, June 1909, *Battiscombe* 19!
TANZANIA. Arusha District: Ngongongare Forest Reserve, Aug. 1951, *Greenway & Hughes* 8563!; Pare District: Mpepera, July 1955, *Semsei* 2094!; Morogoro District: without locality, Aug. 1952, *Semsei* 861!
DISTR. U2, 3; K1, 3–6; T2, 3, 6, 7; Rhodesia, Malawi
HAB. Upland rain-forest; (1000–)1125–2440 m.

SYN. *?C. chirindensis* Engl., V.E. 3 (2): 590, in text (1921), *nom. seminud.* Type: Rhodesia, Melsetter District, Chirinda Forest, *Engler* (B †)
*Rinorea cafassi* Chiov., Racc. Bot. Miss. Consol. Kenya: 6 (1935). Type: Kenya, Mt. Aberdare, *Balbo* 162 (TOM, holo., Kew Negative 4994!)

NOTE. Timber soft, white, easily workable, difficult to season.

## 17. BIVINIA

Tul. in Ann. Sci. Nat., sér. 4, 8: 78 (1857)

Tree. Leaves ± deciduous, alternate, petiolate, penninerved, exstipulate. Flowers bisexual, in axillary racemes; bracts linear; pedicels articulate near the base. Receptacle very shallow. Sepals 5 or 6, valvate. Petals 0. Disk-glands broad and truncate, pubescent, adnate to the base of each sepal. Stamens in fascicles of 10, alternating with the sepals; filaments folded in bud; anthers very small, globose. Ovary 1-locular, with 4–6 multi-ovulate placentas; styles 4–6, filiform. Capsule dehiscent with 4–6 valves. Seeds few, with long hairs; endosperm fleshy.

A monotypic genus in eastern Africa and Madagascar.

**B. jalbertii** *Tul.* in Ann. Sci. Nat., sér. 4, 8: 78 (1857); Mast. in F.T.A. 2: 496 (1871); Perrier in Fl. Madag., Fam. 140: 67, t. 12/1–4 (1946); T.T.C.L.: 548 (1949); Wild in F.Z. 1: 296, t. 53/B (1960); K.T.S.: 497 (1961) as "*jaubertii*". Types: Madagascar, *Richard* 542, *Boivin* 2126, 2567 & *Pervillé* 382 (all P, syn.!)

Shrub or generally a tree, rarely up to 20 m. tall; bark smooth, light grey. Branchlets greyish brown, with pale lenticels, tips pubescent. Leaf-blades ovate, broad-elliptic or subobovate, apex subcaudate-acuminate, base broadly cuneate, thinly chartaceous, crenate except near the base and acumen, slightly pubescent particularly at the midrib and nerves, glabrescent with age, 4·5–10(–13) cm. long, 2·5–4(–6) cm. broad; lateral nerves 7–8 pairs, rather straightly ascending, raised beneath, venation reticulate, but slightly prominent beneath; petiole bright red, pubescent initially, 0·5–1·4 cm. long. Racemes cylindric, densely flowered, 5–12 cm. long, on a short peduncle, covered with a short greyish pubescence; pedicels slender, 2–3(–4) mm. long. Sepals ovate-deltoid, subacute, pubescent on both faces,

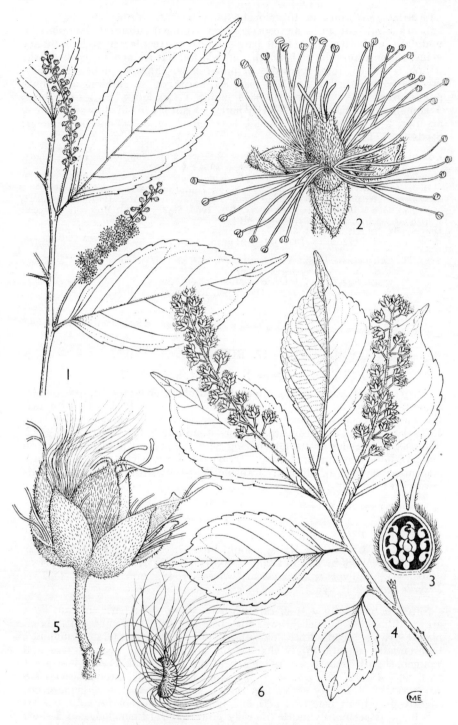

FIG. 17. *BIVINIA JALBERTII*—**1,** flowering branch, × ⅔; **2,** flower, × 8; **3,** longitudinal section of ovary, × 16; **4,** fruiting branchlet, × ⅔; **5,** fruit, × 8; **6,** seed, × 6. 1–3, from *Vaughan* 2727; 4–6, from *Muller & Gordon* 1301. Drawn by Mrs. M. E. Church.

2–3 mm. long, 1·5–2 mm. broad. Glands densely pubescent, truncate, subentire. Filaments slender, glabrous, 3–4 mm. long; anthers 0·2 mm. in diameter. Ovary globose, white tomentulose; styles divergent, glabrous. Capsule globose, pubescent, valves apiculate with the persistent styles. Seeds dark brown, almost cylindric, ± 2 by 1 mm., covered with white cottony hair up to 4 mm. long. Fig. 17.

KENYA. Kwale District: Diani Forest, July 1972, *Gillett & Kibuwa* 19875!; Lamu
    District: Utwani Forest, Jan. 1957, *Rawlins* 321! & Mambosasa Forest, Feb. 1929,
    *R. M. Graham* in *F.D.* 1794
TANZANIA. Uzaramo District: Pugu Hills, Jan. 1939, *Vaughan* 2727!; Rufiji District:
    Mafia I., Aug. 1937, *Greenway* 5098!; Mikindani District: Ruvuma R., Mar. 1861,
    *Kirk*!; Pemba I., Oct. 1929, *Burtt Davy* 22516!
DISTR.   K7; T6, 8; P; also in Rhodesia, Mozambique and Madagascar
HAB.   In open forest and coastal bushland; up to 200 m.

SYN.   *Calantica jalbertii* (Tul.) Warb. in E. & P. Pf. III. 6a: 37, fig. 13/F, G (1893);
    Engl., P.O.A. C: 279 (1895) & V.E. 3 (2): 581, fig. 256/F, G (1921); Gilg in
    E. & P. Pf., ed. 2, 21: 429, fig. 191/F, G (1925)


## 18. **LUDIA**

Juss., Gen.: 343 (1789); Sleumer in Adansonia, sér. 2, 12: 79 (1972)

Trees or shrubs. Leaves persistent, alternate, coriaceous, entire, penninerved, the nerves often steeply ascending, reticulation markedly raised on both faces, petioled, exstipulate. Flowers bisexual, small, solitary, or rarely 2 or 3 together from the same axil, subsessile, subtended by several minute imbricate suborbicular bracts. Sepals 5 or 6, uniseriate, imbricate. Petals 0. Stamens indefinite in number, inserted on a flat receptacle, with numerous hairs between the filiform filaments; anthers small; connective generally rather obscure. Disk-glands several, small, along the edge of the receptacle. Ovary with 2–4 placentas which bear few to numerous ovules; style columnar, often ± deeply 3-lobed or -partite distally. Fruit baccate, hardly or tardily and irregularly dehiscent; pericarp coriaceous. Seeds few; testa hard.

A genus of about 23 species, mainly in Madagascar and the Mascarenes, one also represented in eastern Africa.


**L. mauritiana** *Gmelin*, Syst. Nat., ed. 13, 1: 839 (1791); Sleumer in Adansonia, sér. 2, 12: 100 (1972). Type: Mauritius, *Commerson* (P, lecto.!, W, isolecto.!)

Shrub or small tree, rarely up to 15 m. tall, unarmed. Branchlets early greyish corticate and covered with numerous pale elongate lenticels. Leaf-blades rather variable in shape and size, oblong or oblanceolate, or obovate, or sometimes elliptic, apex usually shortly attenuate and obtuse, even rounded or a little emarginate, base cuneate, coriaceous, shining green above even when dry, entire though with a few distant impressed glands along the lower half of the edge, 3–9 cm. long, 2–4(–6) cm. broad; nerves 6–8 pairs, steeply ascending, not much different from the veins, the latter forming with the veinlets a dense network of elongate meshes which is markedly raised on both faces; petiole 2–8 mm. long. Flowers whitish-yellowish, 1–3 per axil; pedicel 0·5–1 mm. long at anthesis. Sepals ovate-suborbicular, densely pubescent outside and in the upper part of the inside, 2(–3) mm. long, reflexed at anthesis. Disk-glands 10–15, rather small. Stamens 40–60. Ovary glabrous. Fruit subovoid-globular, finely warty, reddish, 1–1·5 cm. across. Seeds 3–8(–12), ± 2 mm. across. Fig. 18.

4

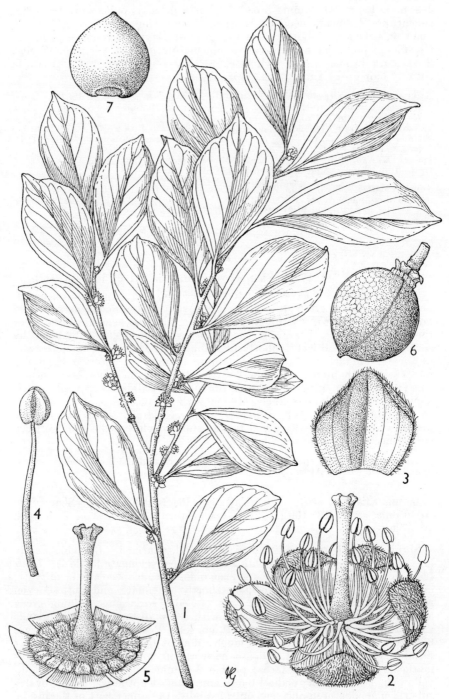

Fig. 18. *LUDIA MAURITIANA*—**1**, habit, × ⅔; **2**, flower, × 10; **3**, sepal, × 12; **4**, stamen, × 18; **5**, pistil, × 12; **6**, fruit, × 2; **7**, seed, × 4. 1–5, from *Faulkner* 2381; 6, from *Faulkner* 2503; 7, from *Faulkner* 2247. Drawn by Victoria Goaman.

KENYA. Nairobi District: Muthaiga, July 1927, *Piemeisel & Kephart* 37!; Kwale District: Jadini, Dec. 1959, *Greenway* 9627!; Kilifi District: Mida, Apr. 1938, *Dale* in *F.D.* 3864!

TANZANIA. Masai/Mbulu District: Lake Manyara National Park, Nov. 1962, *Dingle* 376!; Lushoto District: Sigi, July 1953, *Drummond & Hemsley* 3488!; Pangani District: Boza, Nov. 1956, *Tanner* 3329!; Zanzibar I., Mazizini [Massazine], Mar. 1960, *Faulkner* 2503!

DISTR. **K**4, 7; **T**2, 3, 6, 8; **Z**; Mozambique, Madagascar, Mauritius, Réunion, Seychelles, Aldabra

HAB. Dry evergreen forest and coastal bushland; 0–1750 m.

SYN. *L. sessiliflora* Lam., Encycl. Méth. Bot. 3: 613 (1792); Oliv. in F.T.A. 1: 120 (1868); K.T.S.: 226 (1961). Type: Mauritius, *Commerson* (P, lecto.!, W, isolecto.!)
*Scolopia minutiflora* Sleumer in N.B.G.B. 12: 716 (1935). Type: Tanzania, Lindi District, Mlinguru, *Schlieben* 5756 (B, holo. †, BM, M, P, Z, iso.!)

## 19. **APHLOIA**

(DC.) Benn. in Benn. & Br., Pl. Jav. Rar. 2: 192 (1840)

*Prockia* L. sect. *Aphloia* DC., Prodr. 1: 261 (1824)

*Neumannia* A.Rich., Ess. Fl. Cuba 1: 96 (1845)

Shrubs or trees, entirely glabrous. Leaves persistent, alternate, serrate or serrulate, rarely subentire, penninerved, petiolate; stipules minute, caducous. Flowers bisexual, axillary, solitary or in few-flowered racemes or fascicles, sweet-scented; bracts scale-like, minute; pedicels with 1–3 scaly bracteoles in the lower half. Sepals 4–5(–6), free except at the base, imbricate. Petals 0. Stamens very numerous, free, inserted towards the edge of a slightly concave receptacle; filaments filiform; anthers small, introrse, dorsifixed near the base. Ovary sessile or shortly stipitate, 1-locular, the one parietal placenta with rather few horizontal ovules in 2 rows; stigma subsessile, large, peltate, with a median furrow. Fruit a fleshy berry with ± 6 discoid seeds; testa crustaceous, white, glossy.

Monotypic—though variable as to vegetative characters and the size of the flowers—in eastern Africa, in Madagascar including the Comores, the Mascarenes and Seychelles.

**A. theiformis** (*Vahl*) *Benn.* in Benn. & Br., Pl. Jav. Rar. 2: 192 (1840); Wild in F.Z. 1: 279, t. 48 (1960); K.T.S.: 223 (1961). Type: Réunion, *Commerson* in *Herb. Thouin* (C, holo.!)

Shrub or generally a slender tree, up to 20 m. tall. Branchlets drooping, brown, longitudinally striate with a stronger line decurrent from a stipular cushion. Leaf-blades narrowly elliptic to elliptic, or obovate-elliptic, or oblanceolate, apex subacute or obtuse, base cuneate, firmly chartaceous to subcoriaceous, often becoming bluish-green when dried, glabrous, serrate to serrulate, often entire towards the base, 3–8 cm. long, 1·2–2·8 cm. broad; lateral nerves 8–10 pairs, rather inconspicuous; petioles up to 3 mm. long. Flowers axillary, solitary or 2–3 (rarely more) in a fascicle or short raceme, 8–10 mm. in diameter; pedicels ± 1 cm. long. Sepals white, turning yellowish, orbicular, ± 5 mm. across, connate for 1–1·5 mm. at the base, the inner 3 more membranous and petaloid. Filaments 4–5 mm. long; anthers orbicular, 0·7 mm. across. Ovary ellipsoid; stigma as wide as the ovary, persistent. Berry globose, white, ± 5 mm. across. Seeds ± 2 mm. across. Fig. 19.

KENYA. Teita District: Ngangao Hill, *H. M. Gardner* in *F.D.* 2953! & Wandanyi Mt. top, Oct. 1938, *Joanna* in *Bally* 8939! & 8 km. NNE. of Ngerenyi, Sept. 1953, *Drummond & Hemsley* 4362!

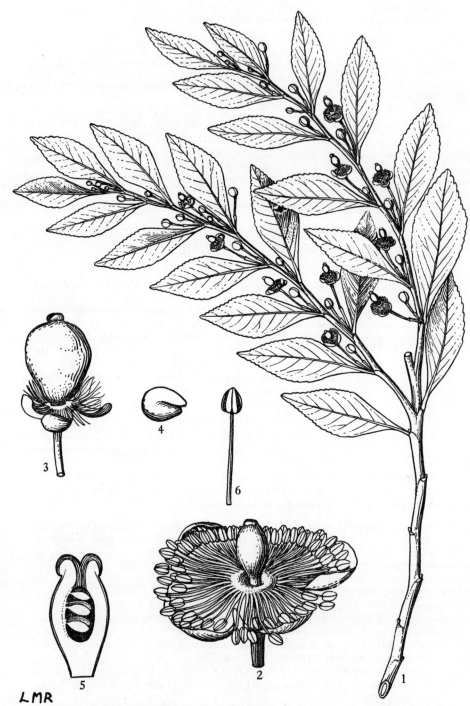

LMR

Fig. 19.  *APHLOIA THEIFORMIS*—**1,** flowering branch, × ⅔; **2,** flower, × 4; **3,** fruit, × 2; **4,** seed, × 4; **5,** longitudinal section of ovary, × 8; **6,** stamen, × 6.  1, from *Johnston* 37; 2, 5, 6, from *Clements* 68; 3, 4, from *Carmichael* 326.  Reproduced by permission of the Editors of Flora Zambesiaca.

TANZANIA. Moshi District: Kilimanjaro, between Umbwe and Weru Weru Rivers, Sept. 1932, *Greenway* 3214!; Morogoro District: Uluguru Mts., Lukwangule, Feb. 1935, *E. M. Bruce* 781!; Songea District: Liwiri-Kiteza Forest Reserve, Oct. 1956, *Semsei* 2556!

DISTR. K7; T2, 3, 6–8; also in Malawi, Mozambique, Rhodesia, South Africa (Transvaal), and in Madagascar including Comores, Mascarenes including Rodriguez I., and the Seychelles

HAB. Upland rain-forest and mist forest, riverine forest, upland evergreen bushland; 1300–2900 m.

SYN. *Lightfootia theiformis* Vahl, Symb. Bot. 3: 69 (1794)
*Prockia theiformis* (Vahl) Willd., Sp. Pl. 2: 1214 (1800)
*Neumannia theiformis* (Vahl) A.Rich., Ess. Fl. Cub.: 97 (1845); Engl., P.O.A. C: 279 (1895) & in E.J. 28: 438 (1900); Gilg in E.J. 30: 360 (1901) & in E.J. 40: 503 (1908); Engl., V.E. 3 (2): 584, fig. 260 (1921); Gilg in E. & P. Pf., ed. 2, 21: 437, fig. 200 (1925)
*Aphloia myrtiflora* Galpin in K.B. 1895: 142 (1895); T.S.K.: 22 (1936); T.T.C.L.: 229 (1949). Type: South Africa, Transvaal, Barberton, summit of Upper Moodies Mountain, *Galpin* 1082 (K, holo.!)
*Neumannia myrtiflora* (Galpin) Th. Dur. & Schinz, Consp. Fl. Afr. 1(2): 218 (1898)

## 20. FLACOURTIA

L'Hérit., Stirp. Nov. 3: 59, t. 30 & 30/B (1786)

Shrubs or trees, sometimes with spiny branches and/or trunk. Leaves alternate, entire or serrate-crenate, penninerved, petiolate, exstipulate. Flowers dioecious, or rarely bisexual, small, in short axillary racemes, sometimes reduced to a solitary flower. Sepals (3–)4–5(–7), slightly connate at the base, imbricate in bud. Petals 0. Male flowers: stamens 15 to numerous, inserted on a receptacle which bears an extra-staminal annular disk usually broken into ± free glands; filaments filiform; anthers dorsifixed. Rudiment of ovary 0. Female flowers: sepals as in the ♂ flowers. Receptacle with an entire, crenulate or lobed disk. Ovary sessile, incompletely (2–)4–6(–8)-locular by false septa; placentas 4–8, pluri-ovulate, with 2 ovules per locule one above the other; styles as many as the locules, free or ± connate, persistent; stigmas small, inflated or shortly 2-lobed. Fruit a fleshy drupe with 4–16 seeds usually in pairs one above the other. Seeds obovoid-ellipsoid, somewhat flattened; testa crustaceous.

A palaeotropical genus with about 10 species, 2 in Africa, one of them in East Africa and extending to tropical Asia and Malesia.

*F. inermis* Roxb., *F. jangomas* (Lour.) Räuschel and *F. rukam* Zoll. & Miq., all native of Asia and/or Malesia, are introduced for their pleasantly edible fruits and are occasionally cultivated within East Africa.

**F. indica** (*Burm. f.*) *Merrill*, Interpret. Rumph. Herb. Amboina: 377 (1917); Gilg in E. & P. Pf., ed. 2, 21: 440, fig. 201 (1925); T.T.C.L.: 231 (1949); I.T.U., ed. 2: 148 (1952); Palgrave, Trees Centr. Afr.: 189, photo. & t. (1957); Wild in F.Z. 1: 286, t. 47/B (1960); K.T.S.: 226 (1961); F.F.N.R.: 265 (1962); Bamps in F.C.B., Flacourt.: 48 (1968); Exell in F.Z. 3: 141, Addenda (1970). Type: Java, *Herb. Burman* (G, holo.!)

Shrub or tree, generally spiny, up to 10 m. tall; bark rough; spines of the trunk sometimes branched, up to 12 cm. long. Vegetative parts varying from glabrous to densely pubescent. Leaves also variable in shape and size; blade ovate or elliptic, sometimes suborbicular or obovate, apex obtusely acuminate, obtuse or rounded, base cuneate to rounded, membranous to almost coriaceous, serrulate-crenate, or more rarely subentire, 2·5–12(–16) cm. long, 2–8 cm. broad; lateral nerves 4–7 pairs, slightly prominent on both faces, as is the ± dense reticulation; petiole up to 2 cm. long.

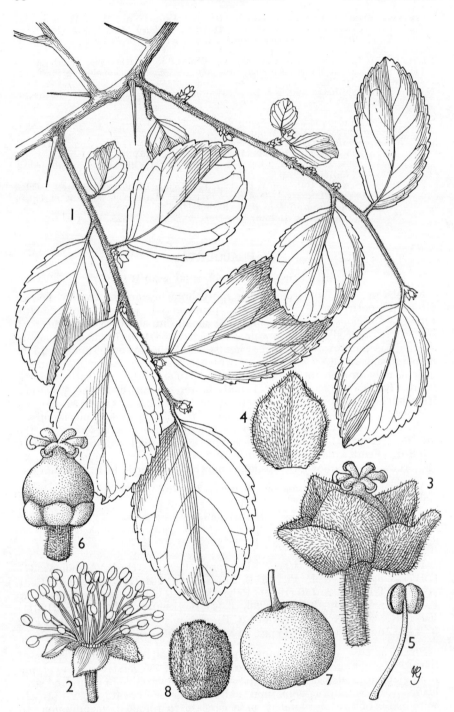

FIG. 20.   *FLACOURTIA INDICA*—**1**, female flowering branch, × ⅔; **2**, male flower, × 8; **3**, female flower,
× 8; **4**, sepal, × 8; **5**, stamen, × 12; **6**, pistil, × 8; **7**, fruit, × 1; **8**, seed, × 3.  1, 3, 4, 6, from *Greenway &
Kanuri* 14749; 2, 5, from *Milne-Redhead & Taylor* 8508; 7, from *Milne-Redhead & Taylor* 10699;
8, from *Boaler* 557.   Drawn by Victoria Goaman.

Flowers dioecious, or occasionally bisexual (1 or several branches of a ♀ specimen with perfect flowers, which, however, bear fewer stamens than in the ♂ ones). Male flowers in axillary racemes 0·5–2 cm. long; pedicels slender, ± pubescent, up to 1 cm. long, the basal bracts minute and caducous. Sepals broadly ovate, apex acute to rounded, pubescent on both sides, 1·5–2·5 mm. long and broad. Filaments 2–2·5 mm. long; anthers 0·5 mm. long. Disk lobulate. Female flowers in short racemes or solitary; pedicels up to 5 mm. Disk lobulate, clasping the base of the ovoid ovary; styles 4–8, central, connate at the base, spreading, up to 1·5 mm. long; stigmas truncate. Fruit globular, reddish to reddish black when ripe, fleshy, up to 2·5 cm. across, with persistent styles, up to 10-seeded. Seeds 8–10 mm. long, 4–7 mm. broad; testa rugose, pale brown. Fig. 20.

UGANDA. W. Nile District: Madi, July 1863, *Speke & Grant*!; Teso District: Serere, Feb. 1933, *Chandler* 1094!; Mengo District: Nakiza Forest, Jan. 1951, *Dawkins* 696!
KENYA. W. Suk District: NW. of Kapenguria, Dec. 1953, *Bogdan* 3886!; Trans-Nzoia District: E. of Kitale, Jan. 1964, *Brunt* 1410!; Lamu District: Pate [Patta] I., *Greenway & Rawlins* 8877!
TANZANIA. Mwanza District: Ukerewe I., Mar. 1929, *Conrads* 5580!; Buha District: Kibondo, Feb. 1955, *Procter* 406!; Lindi District: Rondo Plateau, near Nahoro, Dec. 1955, *Milne-Redhead & Taylor* 7619!; Zanzibar I., Kijunga, Nov. 1930, *Greenway* 2604!
DISTR. U1, 3, 4; K2–5, 7; T1–8; Z; widespread in tropical and subtropical Africa, Madagascar, Mascarenes and Seychelles, also in Asia and Malesia, sometimes cultivated for its edible though acid fruits, or escaped from cultivation
HAB. Woodland, wooded grassland and bushland, often riparian; 0–2400 m.

SYN. *Gmelina indica* Burm. f., Fl. Ind.: 132, t. 39, fig. 5 (1768)
   *Flacourtia ramontchi* L'Hérit., Stirp. Nov. 3: 59, t. 30 & 30/B (1786); Oliv. in F.T.A. 1: 120 (1868); Engl., P.O.A. B: 216, & C: 279 (1895) & in E.J. 28: 439 (1900) & V.E. 3 (2): 585, fig. 261 (1921); T.S.K.: 21 (1936). Type: Madagascar, *L'Héritier* (G, holo.!)
   *F. hirtiuscula* Oliv. in F.T.A. 1: 121 (1868); Engl., P.O.A. B: 217, & C: 279 (1895). Type: Mozambique, near Sena, *Kirk* (K, holo.!)
   *Xylosma ellipticum* Tul. in Ann. Sci. Nat., sér. 5, 9: 343 (1868); Th. Dur. & Schinz, Consp. Fl. Afr. 1(2): 223 (1898). Type: Tanzania, Zanzibar I., *Boivin* (P, holo.!)
   *Flacourtia elliptica* (Tul.) Warb. in E. & P. Pf. III. 6a: 43 (1893); Engl., P.O.A. B: 217, & C: 279 (1895)
   *F. kirkii* Burtt Davy (ined.), Check-Lists Trees & Shrubs Brit. Emp. No. 2, Nyasaland Protect.: 45 (1936), pro syn. *F. hirtiuscula*
   *F. kirkiana* Gardner, T.S.K.: 21 (1936). No type indicated
   *F. afra* Pichi-Serm., Miss. Stud. Lago Tana, 7, Ricerche Bot. 1: 97 (1951); K.T.S.: 225 (1961). Type: Ethiopia, Amhara, Tana, near R. Gueldo (Fissà), *Pichi-Sermolli* 2039 (♂) & 2064 (♀) (FI, holo.!, K, iso.!)

NOTE. The species is conceived here in a wide sense on account of its variability both in Africa and Asia-Malesia. There are, of course, a number of ecotypes which might be locally constant and significant for certain vegetation types.

## 21. DOVYALIS

Arn. in Hook., Journ. Bot. 3: 251 (1841); Gilg in E. & P. Pf., ed. 2, 21: 440 (1925); Sleumer in E.J. 92: 64 (1972)

*Aberia* Hochst. in Flora 27, Bes. Beil.: 2 (1844); Oliv. in F.T.A. 1: 121 (1868)

Shrubs or trees, unarmed, or often with simple or compound spines on trunk and branches, and simple axillary spines on branchlets, dioecious or andromonoecious, i.e. the ♂ plant bearing occasionally a few bisexual flowers and fruits. Leaves alternate, sometimes fascicled on much reduced lateral shoots, generally persistent, rarely deciduous, entire, denticulate or crenate, exstipulate, petiolate, with 1 or 2 basal or suprabasal and a few upper spreading pairs of nerves, with fine pellucid points visible against

strong light in a number of species.  Flowers greenish to yellowish, rarely
white, solitary or in short racemes, these generally reduced to rather few-
flowered fascicles; pedicels with small basal bracts.  Calyx-lobes (3–)4–
6(–7, very rarely –9), free almost to the base, subimbricate to practically
valvate, variously pubescent on both faces, sometimes set with sessile or
stipitate marginal glands, sometimes accrescent in later stages.  Male
flowers: stamens 10–50(–80), inserted on a rather fleshy receptacle, with
several irregularly arranged small disk-glands between; filaments filiform;
anthers dorsifixed.  Rudiment of ovary absent.  Female flowers: calyx-
lobes as many as (rarely –12) and similar to those of the ♂, often slightly
larger, persistent and ± recurved in fruiting time.  Staminodes rarely,
stamens very rarely present.  Ovary surrounded at the base by a cupular
or annular disk (lobed according to the number of the calyx-lobes),
unilocular or incompletely 2–4(–8, rarely more)-locular; placentas 2–4(–8,
rarely more), each with 1 or 2(–6) ovules; styles 2–8(–20, rarely –40),
divergent, channelled, stigmata ± lobed and papillose.  Fruit a fleshy berry.
Seeds few, or occasionally up to 12, ellipsoid, rather flattish, embedded in a
pulp; testa coriaceous, glabrous, hairy, or sometimes woolly.

A genus of about 15 species, 14 confined to continental Africa, one in Ceylon, 6
occurring in East Africa.

D. caffra (Hook. f. & Harv.) Hook. f., native of southern Africa, from Rhodesia to
the Cape Province, is commonly cultivated in the highlands of East Africa, principally
as a hedge plant but also for its edible fruits, as are also to a lesser extent D. hebecarpa
(Gardner) Warb. from Ceylon and D. abyssinica (see below).  The fruits of the East
African species are edible too, but generally of a sour or acid taste.

Ovary with 4–8 placentas, each placenta bearing 2(–6)
    ovules; styles (4–)8–20(–40):
    Seeds glabrous; fruit apple-like, 10–15 cm. across
        when fresh, (5–)6–10 cm. across when dry   .   1. *D. spinosissima*
    Seeds ± densely covered with appressed, subappressed
        or woolly hairs; fruit subglobular, (2–)4–6 cm.
        across when fresh, 2–4 cm. across when dry
        Calyx-lobes of ♂ flowers 5–6 mm. long, 2–3 mm.
           broad; ♀ flowers with 4 or 5 (rarely 6) styles
           per ovary .    .    .    .    .    .   2. *D. abyssinica*
        Calyx-lobes of ♂ flowers 3–4(–5) mm. long, 1–2 mm.
           broad; ♀ flowers with 15–20(–40) styles per
           ovary   .    .    .    .    .    .   3. *D. zenkeri*
Ovary with 2 or 3 (rarely 4) placentas, each placenta
    bearing 1 ovule; styles 2 or 3 (rarely 4):
    Calyx-lobes considerably accrescent (up to 20 mm.
        long and 10 mm. broad), those of the ♂ flowers
        with few or none, those of the ♀ flowers with
        numerous sessile or stipitate marginal glands .   4. *D. macrocalyx*
    Calyx-lobes slightly or not accrescent, and without
        marginal glands:
        Leaves hispidulous on both faces at least initially,
           0·8–3·5 cm. long, 0·6–2·3 cm. broad; ovary
           and fruit practically glabrous (sometimes set
           with a few scattered hispidulous hairs) .    .   5. *D. hispidula*
        Leaves laxly appressed hairy above, densely so
           beneath, 5–8(–11) cm. long, 2·5–5(–7) cm.
           broad; ovary and fruit (even in mature state)
           velvety all over .    .    .    .    .   6. *D. xanthocarpa*

1. **D. spinosissima** *Gilg* in E.J. 40: 509 (1908) & in V.E. 3 (2): 589 (1921)
& in E. & P. Pf., ed. 2, 21: 441 (1925); Wild in F.Z. 1: 282 (1960);
Sleumer in E.J. 92: 66 (1972). Type: Malawi, Blantyre, *Buchanan* 338
(B, holo. †, BM, iso.!)

Much branched shrub or shrubby tree, 2–10 m. tall, often multistemmed
from the base, the distal part of branches pendulous; bark rather rough;
stem with conical (sometimes slightly recurved) bark-covered spines 2(–5)
cm. long. Branchlets slender, brownish to greyish, lenticels scattered, tips
pubescent initially, axillary spines generally present, stoutish, 0·5–6(–7·5)
cm. long. Leaf-blades ovate to elliptic, shortly obtusely acuminate, or
obtuse, base broadly cuneate to rounded, chartaceous, with scattered
pellucid points visible against strong light, sparsely to subdensely appressed
hairy on midrib and nerves on both faces, glabrescent with age, entire or
serrulate, (2·5–)4–9(–18) cm. long, (1·5–)2–5(–8) cm. broad; lateral nerves
4–5 rather steeply ascending pairs, the lowest 1 or 2 pair(s) from the base
or a little above, raised beneath, reticulation obscure; petiole pubescent
initially, 5–10(rarely –20) mm. long. Male flowers 2–5 per fascicle,
tomentellous; pedicels 7–15 mm. Calyx-lobes 5–6(–7), narrowly subacumin-
ate-ovate, sometimes irregularly and minutely laciniate or dented near apex,
entire below, (5–)6–8 mm. long, 2(–4) mm. broad. Disk-glands glabrous or
shortly hairy distally. Female flowers solitary, tomentellous; pedicels
5–10 mm. long. Calyx-lobes 7–10, similar to those of the ♂ flowers, entire,
6–9 mm. long, 2(–4) mm. broad at anthesis, up to 14 mm. long in fruit.
Disk-glands numerous, close together in a kind of ring, glabrous or a little
hairy. Ovary tomentose; styles (7–)8–12(–18), hairy below. Fruit subglobular, a little flattened or deepened on top and base, apple-like, 10–15 cm.
across and greenish yellow when fresh, (5–)6–10 cm. across and becoming
blackish when dry, glabrescent. Seeds rather numerous, glabrous.

UGANDA. W. Nile District: Nyagak R., May 1936, *Eggeling* 3015!; Toro District:
Mpanga R., Oct. 1905, *Dawe* 501!; Kigezi District: Kachwekano near Lake Bunyonyi,
Apr. 1950, *Dawkins* 586!
TANZANIA. Bukoba District: Minziro Forest, Feb. 1958, *Procter* 829!; Arusha District:
Ngurdoto Crater National Park, Momela Gate, Feb. 1966, *Greenway & Kanuri*
12366!; Rungwe District: Masoko, Jan. 1913, *Stolz* 1813!
DISTR. U1, 2; T1, 2, 4, 7; Cameroun, Zaire, Rwanda, Burundi, Malawi
HAB. Openings and margins of rain-forest, dry evergreen and riverine forest, also
secondary growth and bushland; 700–2000 m.

SYN. *D. maliformis* Gilg in V.E. 3 (2): 589 (1921), in key, & in E. & P. Pf., ed. 2, 21:
441 (1925); T.T.C.L.: 231 (1949). Type: Tanzania, Rungwe District, Masoko,
*Stolz* 1813 (B, holo. †, K, iso.!)
*D. sp. nov.* sensu I.T.U.: 148 (1948)
[*D. abyssinica* sensu I.T.U.: 147 (1948), *non* (A.Rich.) Warb.]
*D. macrocarpa* Bamps in B.J.B.B. 34: 500 (1964) & in F.C.B. Flacourt.: 52
(1968). Type: Zaire, Orientale, Nioka, *Bamps* 107 (BR, holo.!)

NOTE. Difficult to distinguish in ♂ and sterile specimens from *D. abyssinica*. In
*D. abyssinica* the petioles are 2–4(–5) mm. long; the spines on the branchlets, if any,
are rather slender and 0·5–1·5, rarely up to 2·5 cm. long. Not known from Kenya yet.


2. **D. abyssinica** (*A.Rich.*) *Warb.* in E. & P. Pf. III. 6a: 44 (1893);
Engl., V.E. 1 (1): 337, fig. 302/D–F (1910); Gilg in V.E. 3 (2): 589,
fig. 262/D–F (1921) & in E. & P. Pf., ed. 2, 21: 441, fig. 203/D–F (1925);
R.E. Fries in N.B.G.B. 9: 325 (1925); T.S.K.: 22 (1936); E.P.A.: 597
(1959); K.T.S.: 224 (1961); Sleumer in E.J. 92: 70 (1972). Type: Ethiopia,
Tigre, Ouodgerate, (*Quartin-Dillon &*) *Petit* 215 (P, holo.!, K, W, iso.!)

Shrub or tree, much branched, up to 8 m. tall; trunk up to 20 cm. across,
with pale greyish-brownish bark. Branchlets sometimes with rather slender

axillary spines 0·5–1·5(rarely –2·5) cm. long.  Leaf-blades ovate-oblong or
-elliptic, apex gradually obtusely attenuate, or sometimes rounded, base
broadly attenuate to obtuse, thinly to more firmly chartaceous, sparsely
and very shortly pubescent on the nerves beneath, glabrescent, entire to
repand-crenate or subserrate, (3–)4–9 cm. long, 2–3·5(–4·5) cm. broad;
lateral nerves 4–5(–7) curved-ascending pairs, slightly raised beneath,
reddish in fresh specimens, no proper reticulation; petiole 2–4(–5) mm.
long.  Male flowers 1–3 in a fascicle in generally already defoliate axils,
rusty-puberulous or tomentellous all over; pedicels 5–8(–13) mm. long.
Calyx-lobes (4–)5(rarely 8), subovate-oblong, subacuminate, 5–6 mm. long,
2–3 mm. broad.  Stamens 40–60.  Interstaminal glands hairy apically.
Female flowers solitary or rarely in pairs in defoliate axils; pedicels
stoutish, 6–8 mm. long.  Calyx-lobes similar to, though slightly larger than
those in the ♂ flowers.  Disk annular, crenulate, finely hairy.  Ovary
rusty-puberulous to -velutinous; styles 4–5(rarely 6).  Fruit globular,
yellowish-reddish with minute paler dots, papillose-puberulous or glabrescent
towards maturity, ± 2 cm. across.  Seeds few, appressed hairy.  Fig.
21/1–9.

UGANDA.  Karamoja District: Napak Mt., June 1950, *Eggeling* 5930 ♂ & 5930 ♀!;
    Mbale District: Mt. Elgon, Bulago, Apr. 1927, *Snowden* 1084!; Mengo District:
    Kampala, Apr. 1917, *Dummer* 3165!
KENYA.  Northern Frontier Province: Mt. Kulal, Oct. 1947, *Bally* 5584!; Meru
    District: Nyambeni Hills, Oct. 1960, *Verdcourt & Polhill* 2966!; Teita Hills, Yale
    Peak, Sept. 1953, *Drummond & Hemsley* 4285!
TANZANIA.  Arusha District: Mt. Meru, Mar. 1970, *Richards* 25587!; Morogoro District:
    Bunduki, Mar. 1953, *Semsei* 1096!; Iringa District: Mt. Image, Mar. 1962, *Polhill &
    Paulo* 1649!
DISTR.  U1, 3, 4; K1–7; T2, 3, 5–7; Ethiopia, Somali Republic, Socotra, Malawi
HAB.  Upland rain-forest to riparian and dry evergreen forest, sometimes in open
    wooded grassland; 1500–3000 m.

SYN.  *Roumea abyssinica* A. Rich., Tent. Fl. Abyss. 1: 34, t. 8 (1847)
    *Dovyalis engleri* Gilg in E.J. 40: 508 (1908); V.E. 1 (1): 337, fig. 302/A–C
        (1910); Gilg in V.E. 3 (2): 589, fig. 262/A–C (1921) & in E. & P. Pf., ed. 2, 21:
        441, fig. 203/A–C (1925); R.E. Fries in N.B.G.B. 9: 326 (1925); T.T.C.L.: 230
        (1949); I.T.U., ed. 2: 147 (1952)

NOTE.  Sterile and ♂ specimens are difficult to distinguish from those of *D. spinosissima*
    (see note there).  The leaves of *D. spinosissima* show irregularly scattered pellucid
    points against strong light, while those of *D. abyssinica* hardly do so.

3. **D. zenkeri** *Gilg* in E.J. 40: 507 (1908); Bamps in F.C.B., Flacourt.:
54, t. 7 (1968); Sleumer in E.J. 92: 72 (1972).  Type: Cameroun, Bipinde,
Bidjoka waterfall, *Zenker* 1543 (B, lecto. †, K, L, M, P, S, isolecto.!)

Shrub or small tree, 1·5–6(–10) m. tall, spiny, with drooping branches.
Branchlets slender, pubescent to hirsute when young, corticate and covered
with numerous lenticels on older parts; spines axillary, slender, up to
5 cm. long.  Leaf-blades ovate to obovate, sometimes ovate-elliptic, apex
shortly even rather abruptly obtusely acuminate, base cuneate to rounded,
or even subcordate, a little oblique, at first thinly papery, more firm when
fully grown, covered with ± appressed substrigose fine hairs on both faces,
more densely or longer so along midrib and nerves, glabrescent with age,
subentire to obscurely serrate, 5–14 cm. long, 2–6 cm. broad; lateral nerves
3–5 pairs, the lowest 1 or 2 pair(s) from the base or a little above,
curved-ascending, raised beneath, reticulation obscure; petiole slender,
5–10 mm. long.  Male flowers 3–4(–6) per fascicle (which is sometimes
shortly stalked), greyish-yellowish velutinous all over; pedicels slender,
3–5(–7) mm. long.  Calyx-lobes (4–)5–7, narrowly ovate-acuminate, 3–4(–5)
mm. long, 1–2 mm. broad.  Stamens 40–50.  Disk-glands small, sparse

FIG. 21. *DOVYALIS ABYSSINICA*—**1,** habit, × ⅔; **2,** male flower, × 4; **3,** female flower, × 4; **4,** sepal, × 8; **5,** stamen, × 12; **6,** pistil, × 6; **7,** transverse section of ovary, × 6; **8,** fruit, × 1; **9,** seed, × 4. *D. MACROCALYX*—**10,** female flower, × 4; **11,** transverse section of ovary, × 12; **12,** young fruit, × 1. 1, from *Shabani* 454; 2, 5, from *Carmichael* 534; 3, 4, 6, 7, from *Drummond & Hemsley* 2798; 8, 9, from *Drummond & Hemsley* 4285; 10, 11, from *Brasnett* 231; 12, from *Osmaston* 1355. Drawn by Victoria Goaman.

between the filaments, pubescent distally. Female flowers solitary; pedicels 4–5 mm. long. Calyx-lobes 7–12, ovate to lanceolate, 4–6(?–10) mm. long, 1·5–2 mm. broad. Disk-glands hairy, close together in the form of a ring. Ovary tomentose; styles 15–20. Hermaphrodite flowers, which occur rarely, as the ♀ ones, but with rather numerous stamens, and only 5 or 6 styles. Fruit subglobular, a little flattened on top and at base, velvety to sparsely pubescent, 4–6(–7·5) cm. across when fresh, 2–3·5 cm. across in the dry state, pale orange yellow. Seeds rather numerous, hairy.

UGANDA. Bunyoro District: Budongo Forest, July 1935, *Eggeling* 2112!; Toro District: Fort Portal Forest Reserve, July 1960, *Paulo* 582!; Kigezi District: Kayonza, Mar. 1947, *Purseglove* 2390!
DISTR. U2; tropical West Africa from Guinea Bissau to the Central African Republic, and in Zaire
HAB. Rain-forest and secondary forest, riverine forest, evergreen bushland; 1500–1675 m.

SYN. *D. afzelii* Gilg in E.J. 40: 507 (1908). Type: Sierra Leone, *Afzelius* (B, holo. †, BM, UPS, iso.!)
    *D. tenuispina* Gilg in Z.A.E.: 570 (1913). Types: Zaire, Kivu Province, Beni, *Mildbraed* 2199 & 2249 (B, syn. †)
    *D. giorgii* De Wild. in B.J.B.B. 4: 408 (1914). Type: Zaire, District Forestier Central, Mobwasa, *De Giorgi* 954 (BR, lecto.!)

4. **D. macrocalyx** (*Oliv.*) *Warb.* in E. & P. Pf. III. 6a: 44 (1893); Gilg in V.E. 3 (2): 588 (1921) & in E. & P. Pf., ed. 2, 21: 441 (1925); T.S.K.: 23 (1936); T.T.C.L.: 231 (1949); F.P.S. 1: 157 (1950); I.T.U., ed. 2: 147 (1952); Wild in F.Z. 1: 283, t. 49/A (1960); K.T.S.: 225 (1961); F.F.N.R.: 265 (1962); Bamps in F.C.B., Flacourt.: 50 (1968); Sleumer in E.J. 92: 75 (1972). Type: Angola, Cuanza Norte, Pungo Andongo, *Welwitsch* 540 (LISU, holo., BM, K, P, iso.!)

Bush or small tree, up to 8 m. tall, much branched, often with a drooping habit; bark smooth. Branchlets slender, tips pubescent, older parts early glabrescent and greyish-brownish corticate, with numerous lenticels, and generally with slender ± straight axillary spines up to 6 cm. long, sometimes unarmed. Leaves subdistichous; blade elliptic or ovate, sometimes narrowly so, apex obtuse or subacute, base broadly cuneate to rounded, rarely subcordate, membranous to chartaceous, glabrous, entire or minutely dentate or remotely crenate, (2·5–)4–9 cm. long, 1·5–4·5 cm. broad, basal or suprabasal lateral nerves 5–7 pairs, steeply ascending, upper 3–5 nerves from the midrib, slightly raised on both faces, reticulation lax, generally rather obscure; petiole 2–5 mm. long. Male flowers solitary or 2–4 per axillary fascicle or abbreviated raceme, tomentellous all over; pedicels 2–3(–6) mm. long. Calyx-lobes 4–6, ovate-lanceolate or lanceolate, subacute, 2·5–4 mm. long, 1(–2) mm. broad, entire or occasionally with a few coarse teeth, sometimes with 1 or 2 gland-hairs. Stamens ± 20. Disk-glands minute, ciliate. Female flowers solitary or 2 (rarely 3) per axillary fascicle, tomentellous all over; pedicels 2–4 mm. long. Calyx-lobes (4–)6–10, lanceolate, often incurved distally, densely ciliate with stalked glands (up to 3 mm. long), 3–6 mm. long and 1–2 mm. broad at anthesis, accrescent to 20 mm. in length and 10 mm. in width at fruiting stage. Annular disk wavy, segmented, tomentellous. Ovary glabrous or laxly hairy; styles 2 (rarely 3). Fruit ellipsoid, fleshy, red to orange, edible, glabrous or laxly hairy, ± 2 cm. long and 1 cm. across, on a pedicel 6–8 mm. long. Seeds 2 (or 3), covered with brownish wool. Fig. 21/10–12, p. 63.

UGANDA. Acholi District: Lomwaga Mt., Apr. 1945, *Greenway & Hummel* 7292!; Busoga District: Butembe Bunya, Dagusi I., Jan. 1953, *G. H. Wood* 605!; Mengo District: Entebbe, Oct. 1905, *Bagshawe* 790!

KENYA. Uasin Gishu District: Eldoret, Apr. 1951, *G. R. Williams* 93!; Central Kavirondo District: Mutet–Kisumu road, Sept. 1953, *Drummond & Hemsley* 4481!; Masai District: Mara R., Apr. 1961, *Glover, Gwynne & Samuel* 402!

TANZANIA. Musoma District: Ikoma, Sept. 1959, *Tanner* 4444!; Mpanda District: Mahali Mts., Sisaga, Aug. 1958, *Newbould & Jefford* 1780!; Bagamoyo District: Kikoka Forest Reserve, Mar. 1964, *Semsei* 3719!; Zanzibar I., *Sacleux* 981!; Pemba I., Oct. 1929, *Burtt Davy* 22572!

DISTR. U1–4; K3–7; T1, 3–7; Z; P; Angola to the Central African Republic, Sudan (Equatoria) and Zaire, Rwanda, Burundi, Mozambique, Malawi, Zambia, Rhodesia

HAB. Rain-forest, dry evergreen and riverine forest, also bushland and wooded grassland; 0–2600 m., but rare below 1200 m.

SYN. *Aberia macrocalyx* Oliv. in F.T.A. 1: 122 (1868)
   *Dovyalis glandulosissima* Gilg in E.J. 40: 506 (1908) & in Z.A.E.: 569 (1913) & in V.E. 3 (2): 588 (1921); T.T.C.L.: 231 (1949); I.T.U., ed. 2: 147 (1952). Type: Tanzania, Iringa District, Uzungwa Mts., Kisinga, *Goetze* 585 (B, holo. †)
   *D. luckii* R.E. Fries in N.B.G.B. 9: 325 (1925). Type: Kenya, Kisumu Londiani District, near Lumbwa, *Fries* 2852 (UPS, holo.!)

NOTE. For other synonyms see Sleumer, *l.c.* (1972).

5. **D. hispidula** *Wild* in Bol. Soc. Brot., sér. 2, 32: 51 (1958) & in F.Z. 1: 284 (1960); Sleumer in E.J. 92: 81 (1972). Type: Rhodesia, Chipinda Gorge, *Davies* 2197 (SRGH, holo.)

Much branched shrub or small tree with spiny stem, up to 6 m. tall; bark grey. Branchlets reddish and patently pubescent at tips, early covered with grey cork and small pale lenticels below; axillary spines slender, rather straight, very acute, up to 4 cm. long and 1–2 mm. across at base. Leaf-blades broad-elliptic to obovate, apex obtuse, rounded or retuse, base cuneate, sometimes obtuse or even subcordate, chartaceous or firmly papery, becoming blackish when dried in young state, otherwise remaining sordidly green, with fine pellucid points visible against strong light, sparsely hispidulous on both faces especially along nerves and edge, the latter shallowly crenate or subentire, 0·8–3·5 cm. long, 0·6–2·3 cm. broad; lateral nerves 1 subbasal pair and 3–4 upper pairs, rather inconspicuous above, visible beneath but fading towards the edge of the lamina, reticulation obscure; petiole hispidulous, very slender, often somewhat curved, 3–6(–8) mm. long. Male flowers solitary or 2–4 per axillary fascicle (these with flowers on top of a very short bracteolate axis), set with white hispid hairs in all outer parts; pedicels very slender, 2–3(–6) mm. long. Calyx-lobes 4–6, ovate, subacute, minutely glandular-dentate in the upper half, ± 2 mm. long. Stamens ± 15. Interstaminal glands sparsely hispidulous distally. Female flowers on short axes similar to the ♂, solitary; pedicels very slender, glabrescent, up to 2 mm. long. Calyx-lobes 5 or 6, broad-ovate, obtuse, often with some glandular minute teeth distally, less hispidulous than in the ♂ ones, ± 2 mm. long and broad, not accrescent. Annular disk subentire, sparsely hispidulous as is the ovary, the latter early glabrescent, or quite often glabrous shortly after anthesis; styles 2 or 3. Fruit subglobular, practically glabrous, possibly still green at maturity, 1–2 cm. across. Seeds 2 or 3, woolly.

KENYA. Kilifi District: Malindi–Kilifi road, Sept. 1936, *Swynnerton* in E.A.H. 59/36; Lamu District: Utwani Forest, Dec. 1956, *Rawlins* 240!

TANZANIA. Tanga District: Siga Caves, June 1918, *Peter* 23793!; Uzaramo District: Fungoni Forest Reserve, Oct. 1965, *Mgaza* 715!; Zanzibar I., Dec. 1930, *Greenway* 2644!

DISTR. K7; T3, 6, 7; Z; Rhodesia, Mozambique

HAB. In riverine thickets and *Brachystegia* woodland, also on limestone and coral rocks; below 150 m.

6. **D. xanthocarpa** *Bullock* in K.B.: 469 (1936); T.T.C.L.: 231 (1949); Sleumer in E.J. 92: 86 (1972). Type: Tanzania, Mpwapwa, Kikombo, *B. D. Burtt* 5006 (K, holo.!, BM, BR, EA, iso.!)

Spreading, much branched, often multistemmed shrub or tree up to 10 m. tall; bark dark grey to brownish, smooth or longitudinally fissured. Branchlets patently hairy at tips, older parts glabrescent, covered with a thin cork and sparse lenticels, sometimes with axillary straight spines up to 2 cm. long and 2 mm. across at base. Leaves deciduous; blade ovate to elliptic, sometimes ± obovate, shortly attenuate and obtuse or subacute at apex, base rounded-truncate, rarely subcordate, thinly chartaceous, becoming blackish when dried, entire, laxly hairy above, densely so or sub-appressed tomentulose beneath especially along midrib and nerves, 5–8(–11) cm. long, 2·5–5(–7) cm. broad; lateral nerves 2 basal or slightly suprabasal, and 3–4 upper pairs, all curved-ascending and becoming more faint towards the edge, slightly raised beneath, reticulation rather inconspicuous; petiole 6–10 mm. long. Male flowers 3–5 per abbreviated cyme-like raceme or fascicle, these mainly from the axils of fallen leaves of second year branch-lets, whitish to yellowish tomentellous all over, on peduncles ± 2 mm. long; pedicels slender, 3–6 mm. long. Calyx-lobes (6–)8, lanceolate, ± 3 mm. long, 1·5–2 mm. broad. Stamens 25–30. Disk-glands very small, globular, extrastaminal or almost so, i.e. about 2 before each calyx-lobe, pubescent. Female flowers solitary, on a short peduncle (1–3 mm. long in anthesis, accrescent to 10 mm. in length in fruit), articulated with a slender pedicel (10–15 mm. long in anthesis, 15–25 mm. long in fruit), grey tomentellous all over. Calyx-lobes 6–8, lanceolate, acuminate, 6–7 mm. long and 1·5 mm. broad, accrescent to 12 × 2–2·5(rarely –4) mm. in fruit. Disk-glands forming a low hairy indistinctly lobed ring. Ovary tomentose; styles 2. Fruit ellipsoid, velvety, orange-yellow, edible, ± 2 cm. long and 1·3 cm. across. Seeds 2, woolly.

TANZANIA. Mbulu District: Lake Manyara National Park, Oct. 1961, *Greenway* 10291!; Handeni District: Misufini, July 1950, *Semsei* 616!; Iringa District: road to Kilosa, near Mtandika, Oct. 1936, *B. D. Burtt* 6072!
DISTR. **T**2, 3, 5–7; not known elsewhere
HAB. In riverine and swampy forest or bushland, sometimes in open bushland; 630–1370 m.

7. **D. sp. A**

Small spiny tree. Leaves ? deciduous; blade elliptic-oblong, laxly short-hairy, ± 3 cm. long, 1–1·5 cm. broad, papery. Calyx-lobes below fruit 4–6, lanceolate, tomentellous, ± 4 mm. long, 1 mm. broad. Fruit globular, dark brown velvety, ± 1 cm. across when dry, said to be edible; styles 2. Seeds 2 or 3, woolly.

KENYA. Kwale District: Mrima Hill, June 1970, *Faden* 70/244; Kilifi District: Arabuko Forest Reserve, Apr. 1930, *Donald* in *F.D.* 2324! & Gedi, Mar. 1954, *Trump* 109!
DISTR. **K**7; not known elsewhere
HAB. In *Afzelia-Mimusops* forest; low altitudes

NOTE. Imperfectly known, should be observed and collected again more completely.

# INDEX TO FLACOURTIACEAE

*Aberia* Hochst., 59
*Aberia macrocalyx* Oliv., 65
**Aphloia** (*DC.*) *Benn.*, 55
*Aphloia myrtiflora* Galpin, 57
**Aphloia theiformis** (*Vahl*) *Benn.*, 55


**Bivinia** *Tul.*, 51
**Bivinia jalbertii** *Tul.*, 51
Bivinia "jaubertii" auct., 51
**Buchnerodendron** *Gürke*, 13
*Buchnerodendron bussei* Gilg, 15
*Buchnerodendron eximium* (Gilg) Engl., 15
**Buchnerodendron lasiocalyx** (*Oliv.*) *Gilg*, 13
*Buchnerodendron nanum* Gilg, 15
*Buchnerodendron stipulatum* (Oliv.) Bullock, 29


*Calantica jalbertii* (Tul.) Warb., 53
**Caloncoba** *Gilg*, 21
*Caloncoba cauliflora* Sleumer, 22
**Caloncoba crepiniana** (*De Wild. & Th. Dur.*) *Gilg*, 24
*Caloncoba gigantocarpa* Perkins & Gilg, 22
*Caloncoba grotei* Gilg, 22
*Caloncoba schweinfurthii* Gilg, 24
**Caloncoba welwitschii** (*Oliv.*) *Gilg*, 22
**Casearia** *Jacq.*, 47
**Casearia battiscombei** *R.E. Fries*, 49
? *Casearia chirindensis* Engl., 51
**Casearia engleri** *Gilg*, 48
*Casearia engleri* sensu auctt., 48, 49
**Casearia gladiiformis** *Mast.*, 49
*Casearia holtzii* Gilg, 49
*Casearia junodii* Schinz, 49
*Casearia macrodendron* Gilg, 49
**Casearia runssorica** *Gilg*, 48
*Chlanis macrophylla* Klotzsch, 21
*Chlanis tettensis* Klotzsch, 19


**Dasylepis** *Oliv.*, 6
*Dasylepis burtt-davyi* Edlin, 6
**Dasylepis eggelingii** *Gillett*, 7
**Dasylepis integra** *Warb.*, 9
*Dasylepis lebrunii* Evrard, 7
*Dasylepis leptophylla* Gilg, 9
*Dasylepis leptophylla* sensu auct., 9
**Dasylepis racemosa** *Oliv.*, 7
*Dasylepis sp.* sensu auct., 9
**Dovyalis** *Arn.*, 59
**Dovyalis abyssinica** (*A. Rich.*) *Warb.*, 61, 60
*Dovyalis abyssinica* sensu auct., 61
*Dovyalis afzelii* Gilg, 64
Dovyalis caffra (Hook. f. & Harv.) Hook. f., 60
*Dovyalis engleri* Gilg, 62
*Dovyalis giorgii* De Wild., 64
*Dovyalis glandulosissima* Gilg, 65

Dovyalis hebecarpa (Gardner) Warb., 60
**Dovyalis hispidula** *Wild*, 65
*Dovyalis luckii* R.E. Fries, 65
**Dovyalis macrocalyx** (*Oliv.*) *Warb.*, 64
*Dovyalis macrocarpa* Bamps, 61
*Dovyalis maliformis* Gilg, 61
**Dovyalis sp. A**, 66
**Dovyalis spinosissima** *Gilg*, 61, 62
*Dovyalis sp. nov.* sensu auct., 61
*Dovyalis tenuispina* Gilg, 64
**Dovyalis xanthocarpa** *Bullock*, 66
**Dovyalis zenkeri** *Gilg*, 62


*Eriudaphus zeyheri* Nees, 37


**Flacourtia** *L'Hérit.*, 57
*Flacourtia afra* Pichi-Serm., 59
*Flacourtia elliptica* (Tul.) Warb., 59
*Flacourtia hirtiuscula* Oliv., 59
**Flacourtia indica** (*Burm. f.*) *Merrill*, 57
Flacourtia inermis Roxb., 57
Flacourtia jangomas (Lour.) Räuschel, 57
*Flacourtia kirkiana* Gardner, 59
*Flacourtia kirkii* Burtt Davy, 59
*Flacourtia ramontchi* L'Hérit., 59
Flacourtia rukam Zoll. & Miq., 57


**Gerrardina** *Oliv.*, 37
**Gerrardina eylesiana** *Milne-Redh.*, 39
Gerrardina foliosa Oliv., 39
*Gmelina indica* Burm. f., 59
**Grandidiera** *Jaub.*, 11
**Grandidiera boivinii** *Jaub.*, 11


**Homalium** *Jacq.*, 42
**Homalium abdessammadii** *Aschers. & Schweinf.*, 42
subsp. wildemanianum (Gilg) Wild, 43
**Homalium africanum** (*Hook. f.*) *Benth.*, 46
*Homalium boehmii* Gilg, 43
*Homalium calodendron* Gilg, 46
*Homalium eburneum* Engl., 43
**Homalium elegantulum** *Sleumer*, 43
**Homalium gracilipes** *Sleumer*, 45
**Homalium longistylum** *Mast.*, 45
*Homalium macranthum* Gilg, 43
*Homalium molle* Stapf, 47
var. *rhodesicum* R.E. Fries, 47
*Homalium mossambicense* Paiva, 46
*Homalium rhodesicum* Dunkley, 43
*Homalium riparium* Gilg, 47
*Homalium rufescens* sensu auct., 43
*Homalium sarcopetalum* Pierre, 46
*Homalium setulosum* Gilg, 43
*Homalium stipulaceum* sensu auct., 47
*Homalium stuhlmannii* Warb., 43

*Homalium warburgianum* Gilg, 43
*Homalium wildemanianum* Gilg, 43


**Kiggelaria** *L.*, 31
**Kiggelaria africana** *L.*, 31
*Kiggelaria flavo-velutina* Sleumer, 33
*Kiggelaria glabrata* Gilg, 33
*Kiggelaria grandifolia* Warb., 31
*Kiggelaria hylophila* Gilg, 33
*Kiggelaria serrata* Warb., 31


*Lightfootia theiformis* Vahl, 57
**Lindackeria** *Presl*, 24
*Lindackeria bequaertii* De Wild., 27
**Lindackeria bukobensis** *Gilg*, 25
*Lindackeria dentata* sensu auct., 27
**Lindackeria fragrans** (*Gilg*) *Gilg*, 25
*Lindackeria grewioides* Sleumer, 29
*Lindackeria kivuensis* Bamps, 27
*Lindackeria mildbraedii* Gilg, 27
**Lindackeria schweinfurthii** *Gilg*, 27
*Lindackeria somalensis* Chiov., 27
*Lindackeria sp. aff. grewioides* Sleumer, 27
**Lindackeria stipulata** (*Oliv.*) *Milne-Redh. & Sleumer*, 28
**Ludia** *Juss.*, 53
**Ludia mauritiana** *Gmelin*, 53
*Ludia sessiliflora* Lam., 55


*Neumannia* A. Rich., 55
*Neumannia myrtiflora* (Galpin) Th. Dur. & Schinz, 57
*Neumannia theiformis* (Vahl) A. Rich., 57


**Oncoba** *Forssk.*, 15
*Oncoba angustipetala* De Wild., 21
*Oncoba boivinii* (Jaub.) Baill., 13
*Oncoba crepiniana* De Wild. & Th. Dur., 24
*Oncoba dentata* sensu auct., 27
*Oncoba eximia* Gilg, 15
*Oncoba fissistyla* Warb., 21
*Oncoba fragrans* Gilg, 25
*Oncoba kirkii* Oliv., 21
*Oncoba lasiocalyx* Oliv., 15
*Oncoba macrophylla* (Klotzsch) Warb., 21
*Oncoba micrantha* Gilg, 18
*Oncoba petersiana* Oliv., 21
**Oncoba routledgei** *Sprague*, 16
*Oncoba spinidens* Hiern, 4
**Oncoba spinosa** *Forssk.*, 16
  var. angolensis Oliv., 16
  var. *routledgei* (Sprague) Dale & Greenway, 18
*Oncoba stipulata* Oliv., 29
*Oncoba stuhlmannii* Gürke, 21
*Oncoba tettensis* (Klotzsch) Harv., 19
*Oncoba welwitschii* Oliv., 22


**Peterodendron** *Sleumer*, 9
**Peterodendron ovatum** (*Sleumer*) *Sleumer*, 11

**Phylloclinium** *Baillon*, 29
*Phylloclinium brevipetiolatum* Germain, 29
**Phylloclinium paradoxum** *Baillon*, 29
*Poggea ovata* Sleumer, 11
Prockia L.
  sect. *Aphloia* DC., 55
*Prockia theiformis* (Vahl) Willd., 57


**Rawsonia** *Harv. & Sond.*, 3
**Rawsonia lucida** *Harv. & Sond.*, 4
**Rawsonia reticulata** *Gilg*, 6
*Rawsonia schlechteri* Gilg, 4
· *Rawsonia spinidens* (Hiern) Mendonça & Sleumer, 6
*Rawsonia transjubensis* Chiov., 6
*Rawsonia ugandensis* Dawe & Sprague, 4
*Rawsonia uluguruensis* Sleumer, 6
*Rawsonia usambarensis* Engl. & Gilg, 4
*Rinorea cafassi* Chiov., 51
*Roumea abyssinica* A. Rich., 62


**Scolopia** *Schreb.*, 33, 3
*Scolopia cuneata* Warb., 37
*Scolopia guerkeana* Gilg, 35
*Scolopia minutiflora* Sleumer, 55
**Scolopia rhamniphylla** *Gilg*, 35
Scolopia "rhamnophylla" auct., 35
*Scolopia rigida* R.E. Fries, 37
*Scolopia riparia* Mildbr. & Sleumer, 35
**Scolopia stolzii** *Gilg*, 34
  var. **riparia** (*Mildbr. & Sleumer*) *Sleumer*, 35
  var. **stolzii**, 34
*Scolopia stuhlmannii* Warb. & Gilg, 37
**Scolopia theifolia** *Gilg*, 33
Scolopia "theiformis" auct., 33
*Scolopia zavattarii* Chiov., 34
**Scolopia zeyheri** (*Nees*) *Harv.*, 37


**Trimeria** *Harv.*, 39
*Trimeria bakeri* Gilg, 40
**Trimeria grandifolia** (*Hochst.*) *Warb.*, 40
  subsp. grandifolia, 40
  subsp. **tropica** (*Burkill*) *Sleumer*, 40
*Trimeria macrophylla* Bak. f., 40
*Trimeria tropica* Burkill, 40


*Xylosma ellipticum* Tul., 59
**Xylotheca** *Hochst.*, 18
*Xylotheca fissistyla* (Warb.) Gilg, 21
*Xylotheca glutinosa* Gilg, 21
*Xylotheca holtzii* Gilg, 21
*Xylotheca kirkii* (Oliv.) Gilg, 21
*Xylotheca kraussiana* Hochst., 18
*Xylotheca macrophylla* (Klotzsch) Sleumer, 21
*Xylotheca stuhlmannii* (Gürke) Gilg, 21
*Xylotheca sulcata* Gilg, 21
**Xylotheca tettensis** (*Klotzsch*) *Gilg*, 19
  var. **fissistyla** (*Warb.*) *Sleumer*, 21
  var. **kirkii** (*Oliv.*) *Wild*, 21
  var. **macrophylla** (*Klotzsch*) *Wild*, 19
  var. **tettensis**, 19

NOTE. One new combination, *Xylotheca tettensis* (Klotzsch) Gilg var. *fissistyla* (Warb.) Sleumer, is published on p. 21.